Life Under The Lens

Jennifer Delaney

ISBN: 978-1999742201

Created by Jennifer Delaney

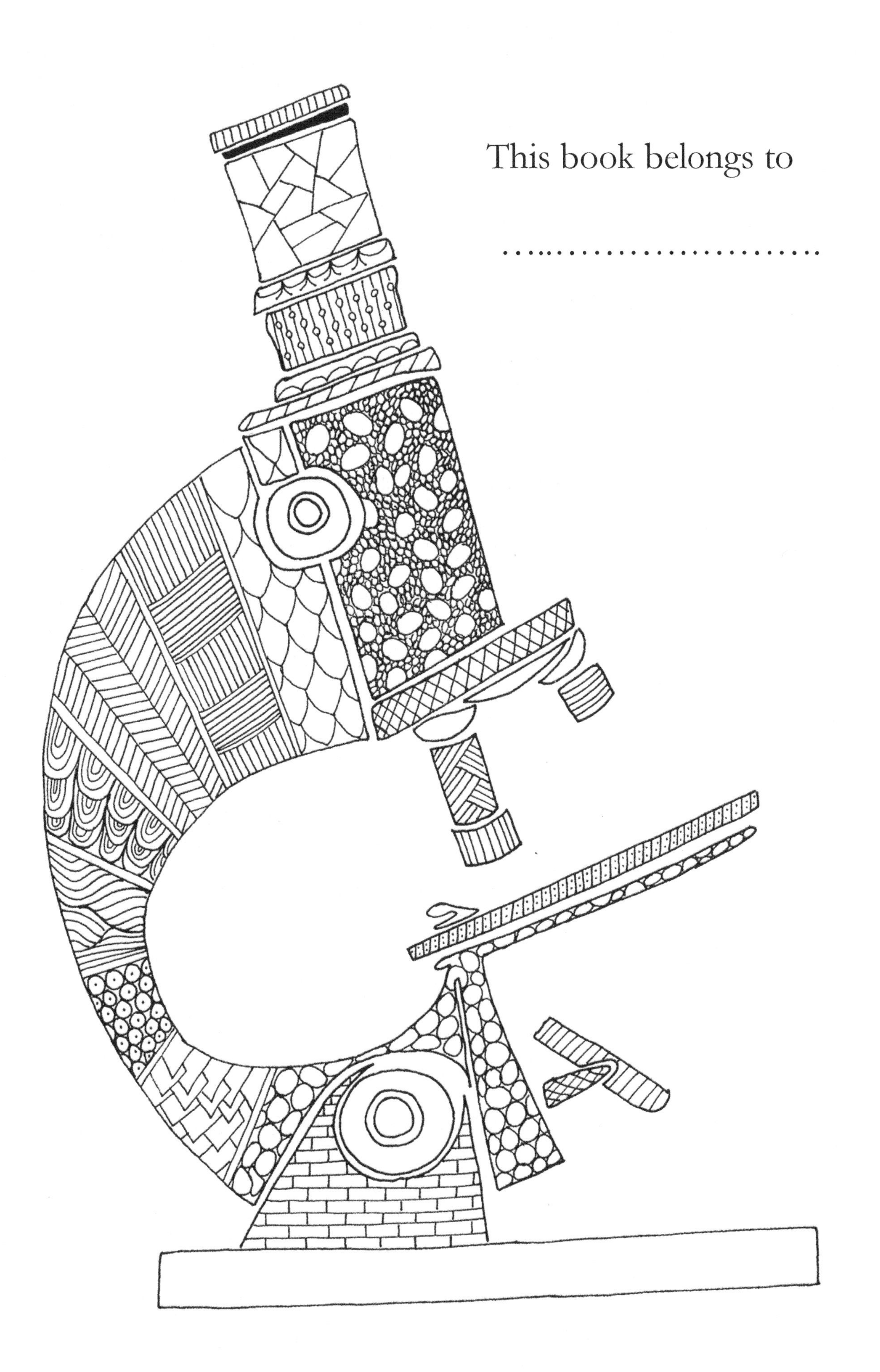

This book belongs to

…..……………………

INTRODUCTION

The relationship between science and art has developed together from a time before their titles were attributed. When our ancestors represented their environment on the walls of their dwellings they done so with scientific observation, artistic expression and creative skill. Through time mathematics became the vehicle used to record experiment and results. Today art is very important in the area of science communication which is used to drive scientific education, public interest and funding. Scientific art also provides a means to investigate our surroundings and ignite our natural curiosity regarding the world in which we live.

This book is both a field guide to the microscopic world and a therapeutic colouring book which aims to illuminate some of the findings of modern science and technology. It brings together a collection of illustrations of common bacteria, ancient archaea, fungi, protists such as radiolarians and foraminifera, cross sections of plants and microscopic animals.

My drawings are stylised and highly decorated for the colouring artist while visually communicating the structure and biological functions of the organisms for the student. Each drawing is accompanied by a short piece of text highlighting important morphological features and other significant details. This unique colouring book brings the user to the unobservable, a world beyond the limits of our own vision.

Archaea

Archaea are single celled organisms and among the earliest forms of life on Earth. They are prokaryotes but are only distantly related to the bacteria. Like bacteria, they too have no nucleus or mitochondria.

Archaea are especially suited to living in harsh environments. Some live near hydrothermal vents in the deep ocean at temperatures over 100° C. They thrive in hot springs or very alkaline or acidic waters. Most gain nourishment from dead material, while others infest living bodies.

For over 4 billion years, prokaryotes have greatly affected our planet's climate, rock formation, and evolution of all other life on Earth.

Bacteria

Bacteria are prokaryotes that are one of the most numerous and diverse organisms on Earth. Bacteria are single celled, and similar to Archaea have no nucleus or mitochondria. They are notorious for causing infectious diseases while many are actually essential to human health.

Bacteria cells have a different physical and chemical make up to Archaea cells, especially in their cell wall. These characteristics give them the ability to live in every environment.

Escherichia coli (E.coli), illustrated here, is a well-known bacteria found in the lower intestine of warm blooded animals. It is the cause of food poisoning in humans although some strains of E-coli are also harmless. Here you can see the rod-shaped bacterium.

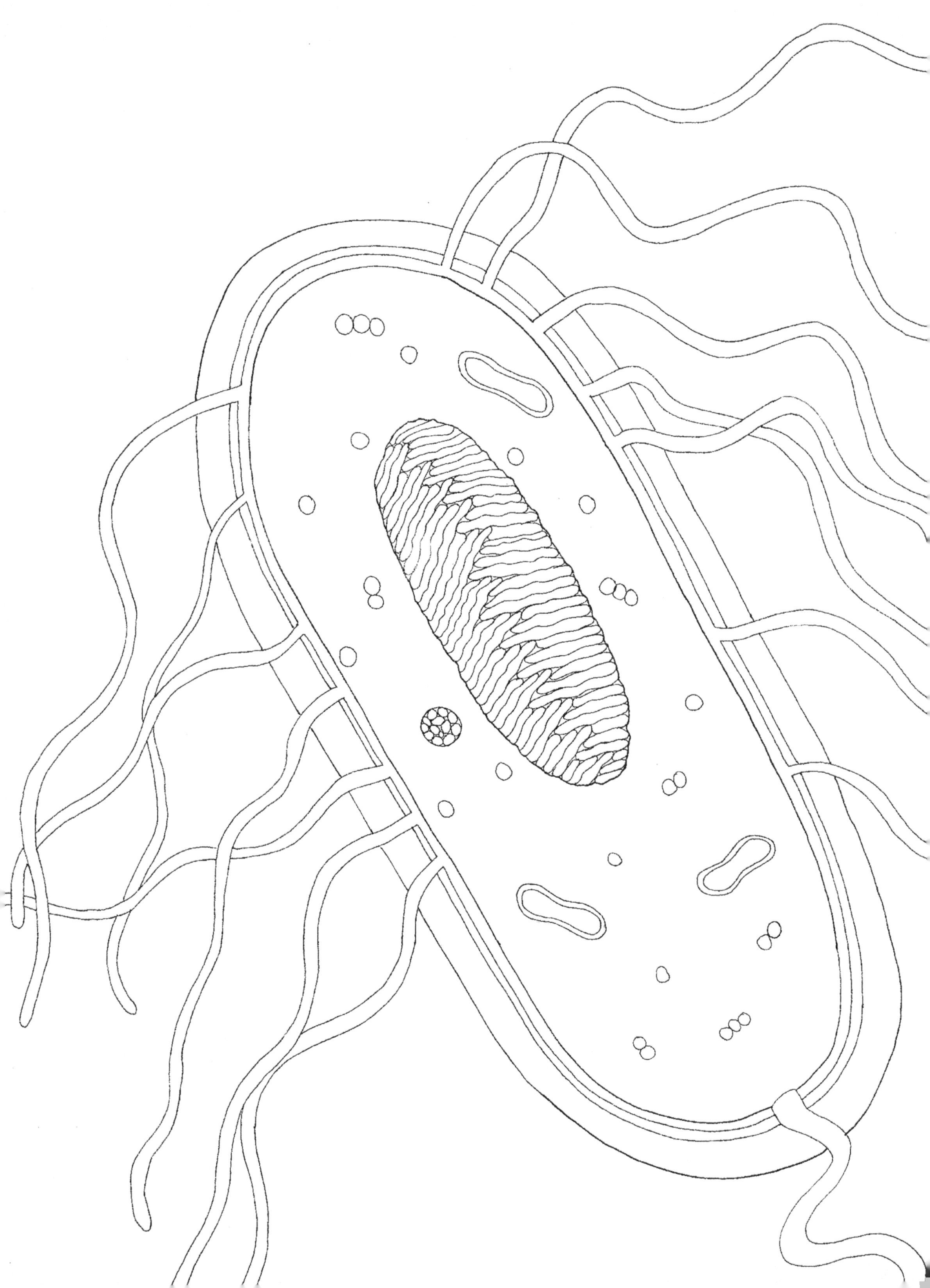

Cyanobacteria

Cyanobacteria are photosynthetic and aquatic bacteria which use the sun's energy and convert it into usable energy via photosynthesis. They contain the oldest known fossils which are around 3.5 billion years old.

Cyanobacteria can occur as single cells, filaments of cells, or colonies. When exposed to phosphates and nitrogen they can produce algal blooms that cover hundreds of square miles. These algal blooms eventually decay, leaving behind a severely hypoxic environment which can be toxic to marine life.

Cyanobacteria are important in shaping the course of evolution. They are credited with converting the Earth's early oxygen poor atmosphere into an oxidizing atmosphere, suitable for aerobic life.

Giardia lamblia

Giardia lamblia is a flagellate. They are swimming microbes what move themselves using whip-like propulsion. They are an intestinal parasite of mammals including humans that cause the disease giardiasis. They enter the human system through ingestion of a dormant cyst.

Giardia have a paired nuclei outlined by adhesive discs which are bilaterally symmetrical.

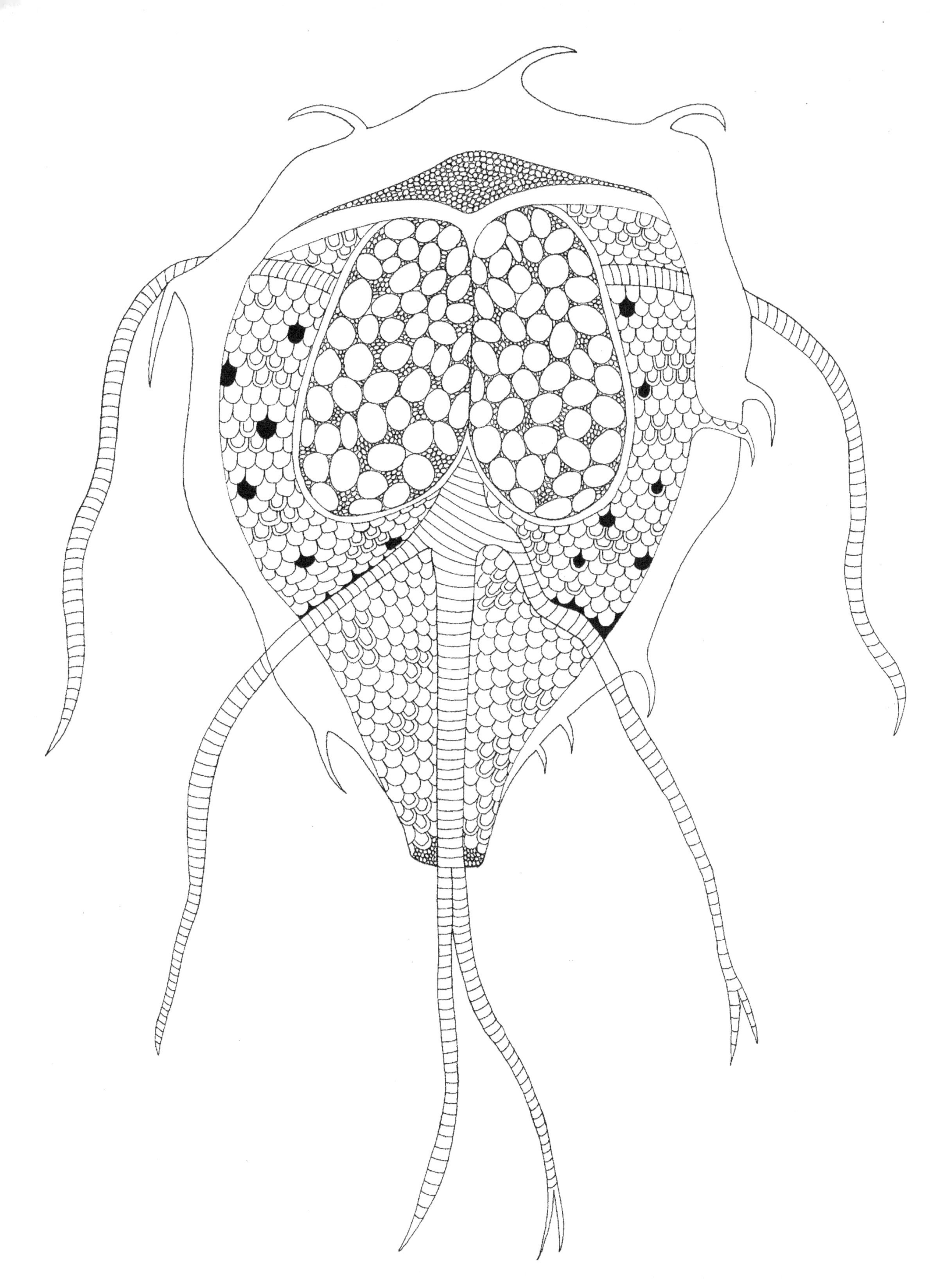

Arcella dentata

Arcella dentata is an amoeba found in bogs and swamps. Arcella species have circular tests made from chiton. They use pseudopods or "false feet" to creep forward and hunt for smaller organisms such as diatoms, unicellular algae and flagellates. They envelope their prey, consuming them while they are still alive.

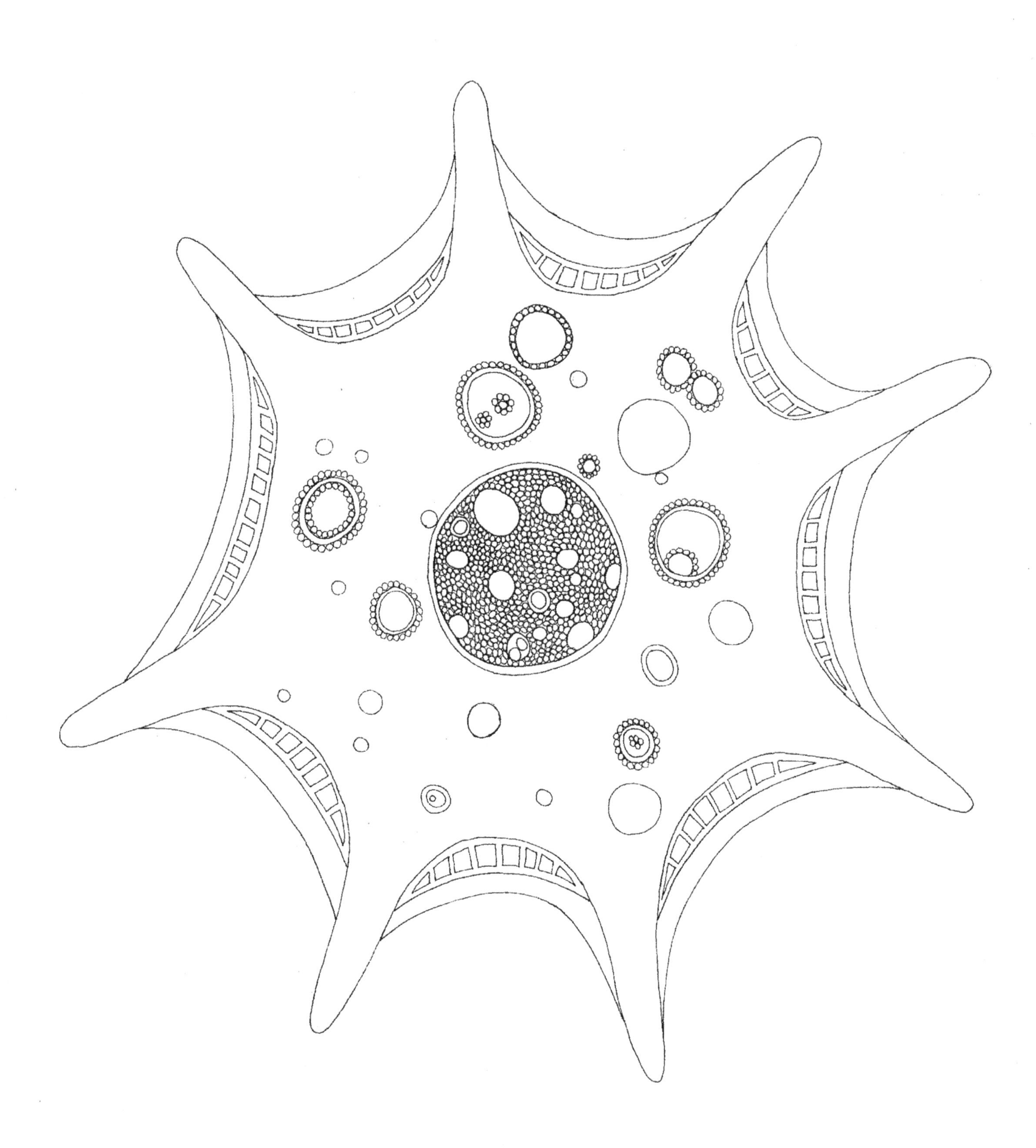

Radiolarians

Radiolarians are marine zooplankton and occur in all oceans, shallow seas, bays and fjords of the world. Most are found in the upper 100 metres of the open oceans but they can be found at all depths. Their fossil record can be traced back to around 600 million years ago.

They are complex and sophisticated organisms with glassy skeletons and often perfect geometry. They exhibit an astonishing variety of intricate shapes and spines. They can be spherical, cone-shaped, rod shaped and possess radial to bilateral symmetry. Their outer skeleton is composed of a crystallised form of opaline silica, which provides insight into their ecology and is used to classify them. Their skeletons are generally organised around spines called spicules, which form dense projections from the main part of the skeleton.

Their soft anatomy which is held inside the central capsule contains the endoplasm and nucleus. It is surrounded by the ectoplasm which has frothy bubble like alveoli which aids in buoyancy. The ectoplasm has a corona of rays which are used in feeding, as well as increasing their surface area for buoyancy.

Radiolaria are non-motile drifters which feed along ocean currents. They consume a large variety of minerals gaining their nutrition from copepods, crustacean larvae, flagellates, ciliates, diatoms, coccolithophores and dinoflagellates. They are known to consume bacteria and detritus also. Some species act as predators and capture prey such as diatoms with their rays called axopodia.

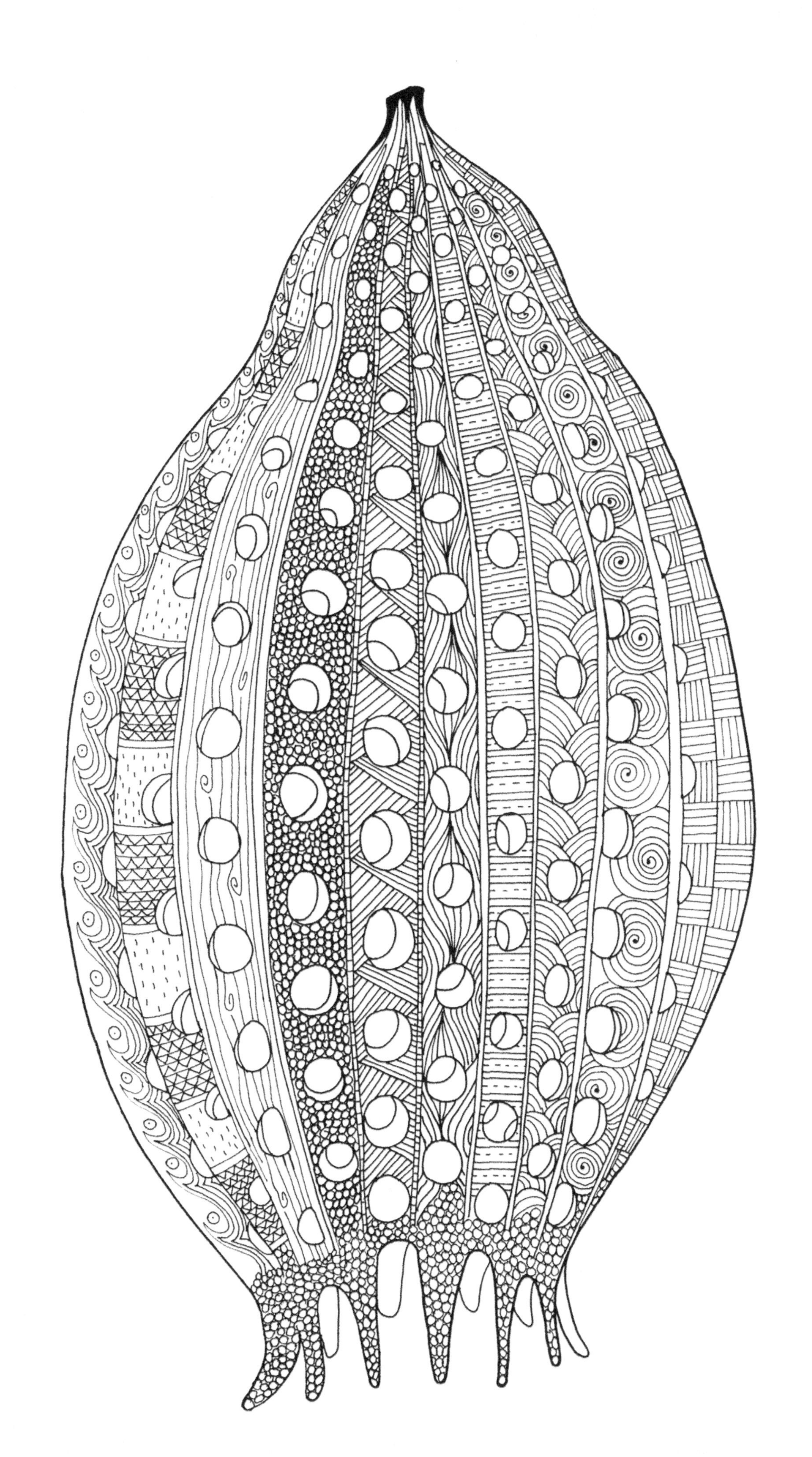

Dinoflagellates

Dinoflagellates are unicellular protists which possess a great diversity of form. They can be photosynthetic, making their own food using the sun's energy or non-photosynthetic feeding on diatoms and other protists. Some species are able to make their own light through a process called bioluminescence.

Each individual had just one cell with two whip-like hairs called flagella. These cells can be naked or may have an outer armor of cellulose plates called thecae.

Dinoflagellates are responsible for "red tides" when they reproduce in great numbers. They produce a neurotoxin which affects humans who consume fish or shellfish containing the toxins.

Dinoflagellates are mainly marine with some freshwater species. They are second only to diatoms as the primary producers of organic matter in the world's oceans.

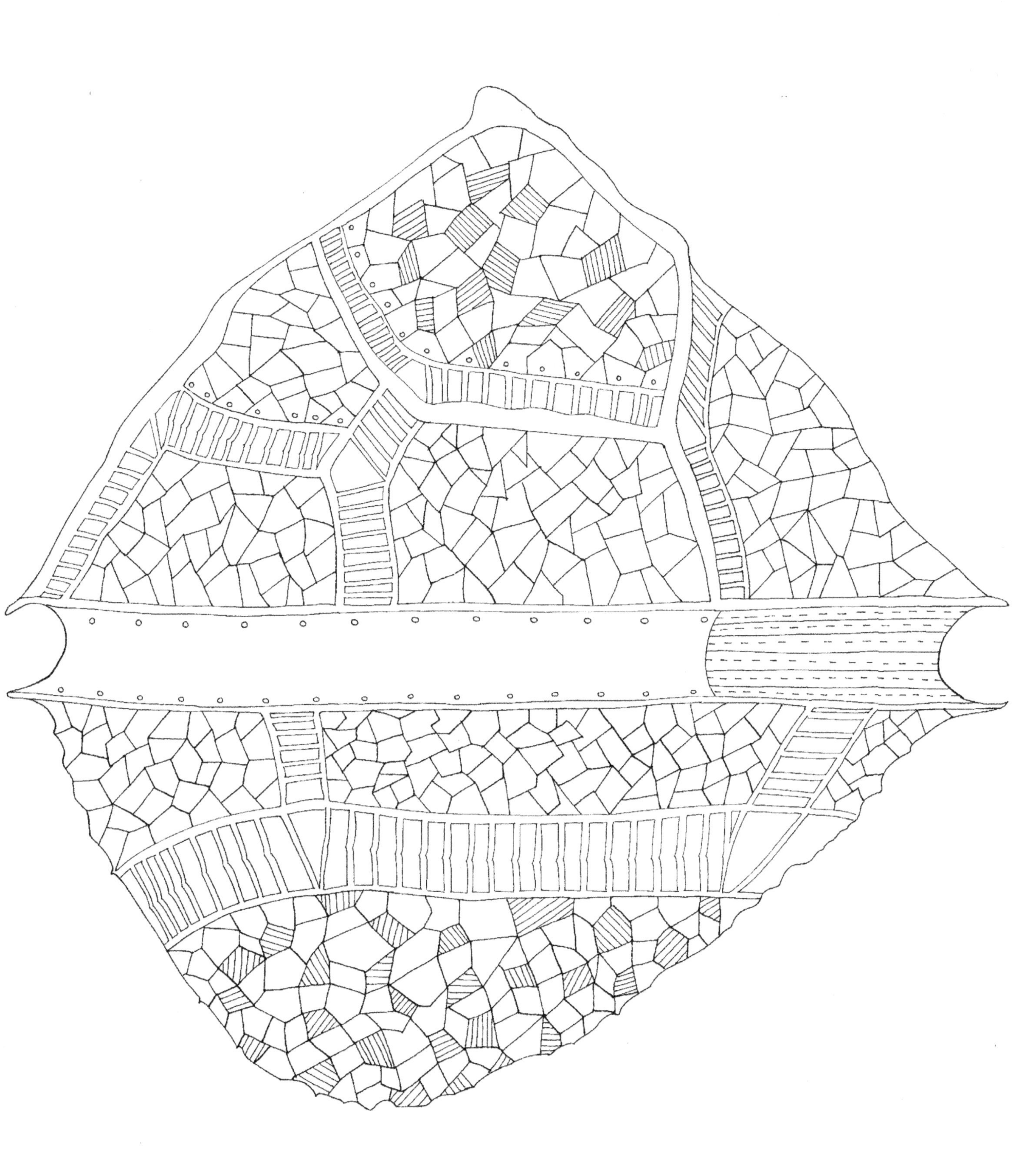

Euglena

Euglena is a genus of flagellate found in both fresh and salt water environments. Most species of Euglena have chloroplasts used for photosynthesis like land plants, but like animals they can gain nourishment by osmosis.

Within the body, Euglena have a nucleus containing the nucleolus, chloroplasts and vacuole. Some species have a short flagellum and one long flagellum, an eye spot and a light sensor.

Euglena lack a rigid cellulose wall and have a flexible pellicle that allows them to change shape.

Foraminiferans

Foraminiferans or forams have flourished in our oceans for hundreds of millions of years. Their calcified shells form layers of chalky sediment on our planet's sea floor. They have left an impressive fossil record.

Within their calcified shells, called tests, are tiny amoeba which hunt their prey with pseudopods. The tests have one or multiple chambers, many possessing elaborate designs made of calcium carbonate.

Diatoms

Diatoms are unicellular algae and are among the most common group of phytoplankton having been on this Earth before the Jurassic period. There are 200 genera of living diatoms inhabiting oceans, freshwater, soils and even damp surfaces. They come in a large variety of shapes and sizes, some forming chain like colonies like filaments or ribbons, while other are shaped like fans, circles, triangles, zigzags and stars. They are usually microscopic, ranging in size from 2 -200 micrometres.

Diatoms are traditionally divided into two categories, centric or pennate. Centric diatoms show radial symmetry whereas pennate diatoms show bilateral symmetry. A long raphe runs down the centre of most pennate diatoms and they have a large central vacuole or pair of vacuoles.

The cell wall is called a frustule and there is huge diversity in frustule form. Their walls are made of silica. Most show bilateral symmetry although one frustule is slightly larger, overlapping the other frustule like a petri dish. The cell walls are highly patterned with pores, ribs, minute spines, marginal ridges and elevations.

When the diatoms decompose they leave their organic and inorganic remains in the sediment thus are well documented in the fossil record.

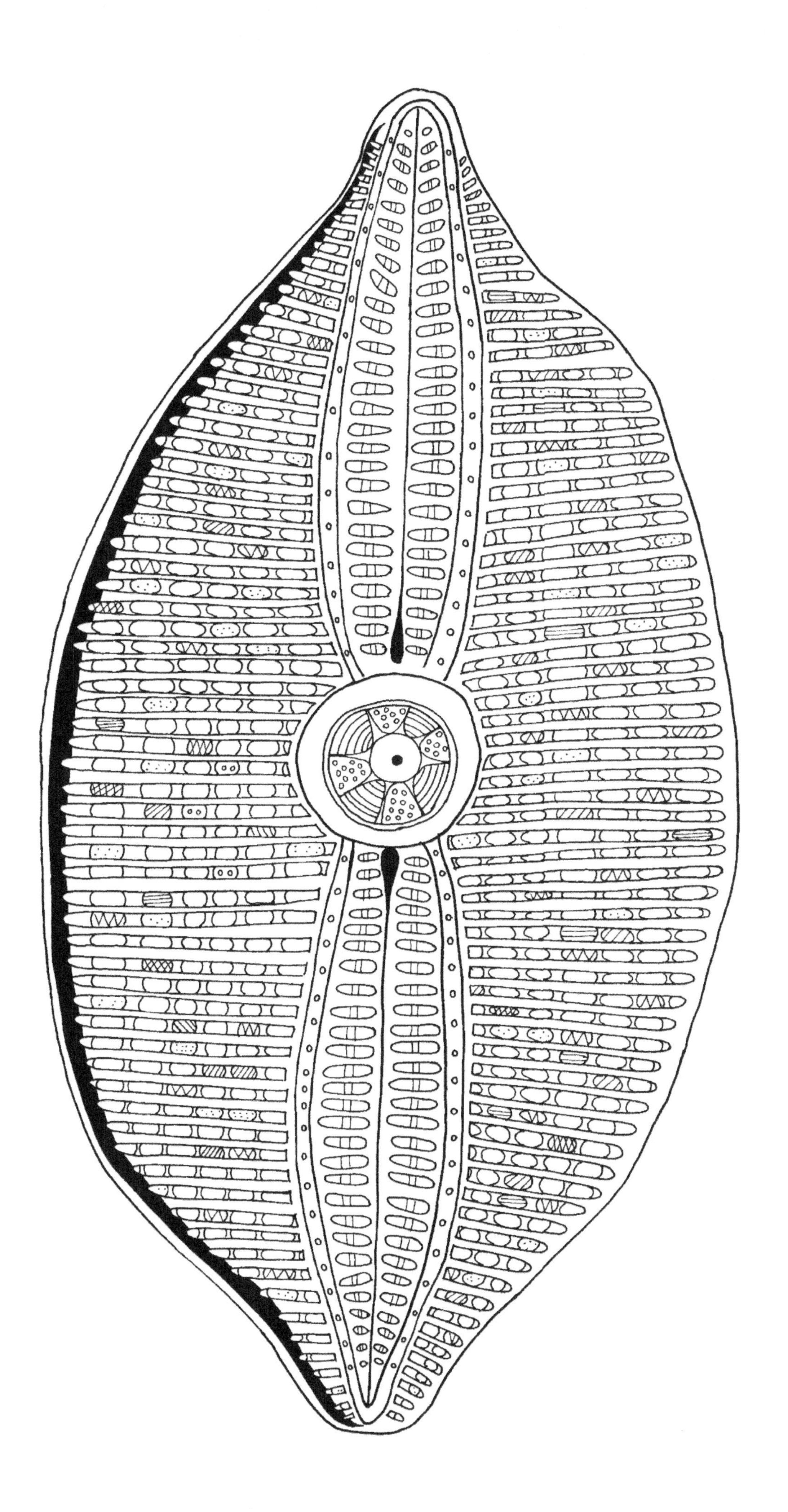

Diatoms

Diatoms reproduce asexually through binary fission where each daughter cell receives one of the parents two frustules, and then grows a second frustule into this.

They are referred to as producers in the food chain, as they make their own food from sunlight and chemical energy. They are estimated to contribute up to 45% of total organic primary production in our oceans. They have a boom or bust type existence. They boom when there is an abundance of nutrients and light and they sink when conditions are unfavourable.

Diatoms are used for monitoring environmental conditions today as well as being indicators of ancient climates.

Red algae

Red algae include microscopic organisms and large multicellular seaweeds. They can live in deep coastal waters as they can absorb blue light waves that other green or brown algae cannot.

Polysiphonia, drawn here, is a filamentous red algae. It has many branches and grows attached to rocks via rhizoids. It consists of a central filament that supports pericentral cells. Each branch is an axis or series of elongated cells which give the algae a segmented appearance.

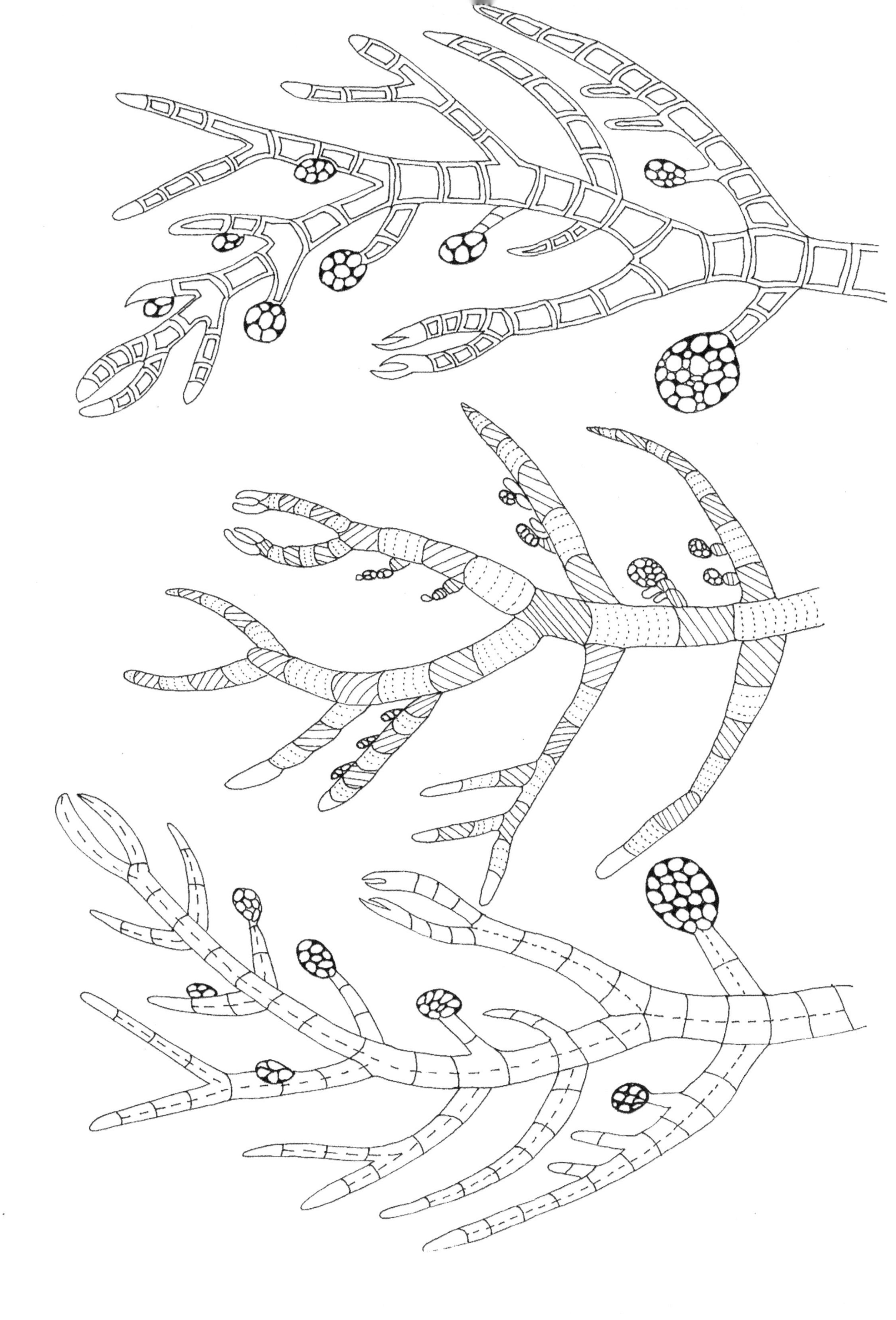

Desmids

Desmids are microscopic plants which are flowerless and have no roots, stems or true leaves. They have the largest single cells in the entire plant world with some being visible to the naked eye. Desmids are related to a group of land plants which includes liverworts, mosses, ferns and flowering plants.

They are unicellular and possess semi-cells which are divided by a central connection, called an isthmus. The nucleus is located within the isthmus. They are characterised by their extensive variation in cell shape. The semi-cells are undulated or are indented with processes. The cell walls can be smooth but they often have definite tubercules, spines or granules. Cells can form attractive elongated, moon shaped, star-shaped and rotund configurations. Some desmids form long filaments which resembles a helix.

Desmids photosynthesise by capturing the sun's energy. They make sugars and starch by utilising the carbon dioxide in the surrounding water. They are found in clear, relatively nutrient poor waters, often between sphagnum moss in bogs and marches.

Desmids play a part in recycling quantities of barium and strontium in these freshwater environments. Desmids are excellent indicator organisms which are useful for monitoring conservation value in semi-aquatic habitats.

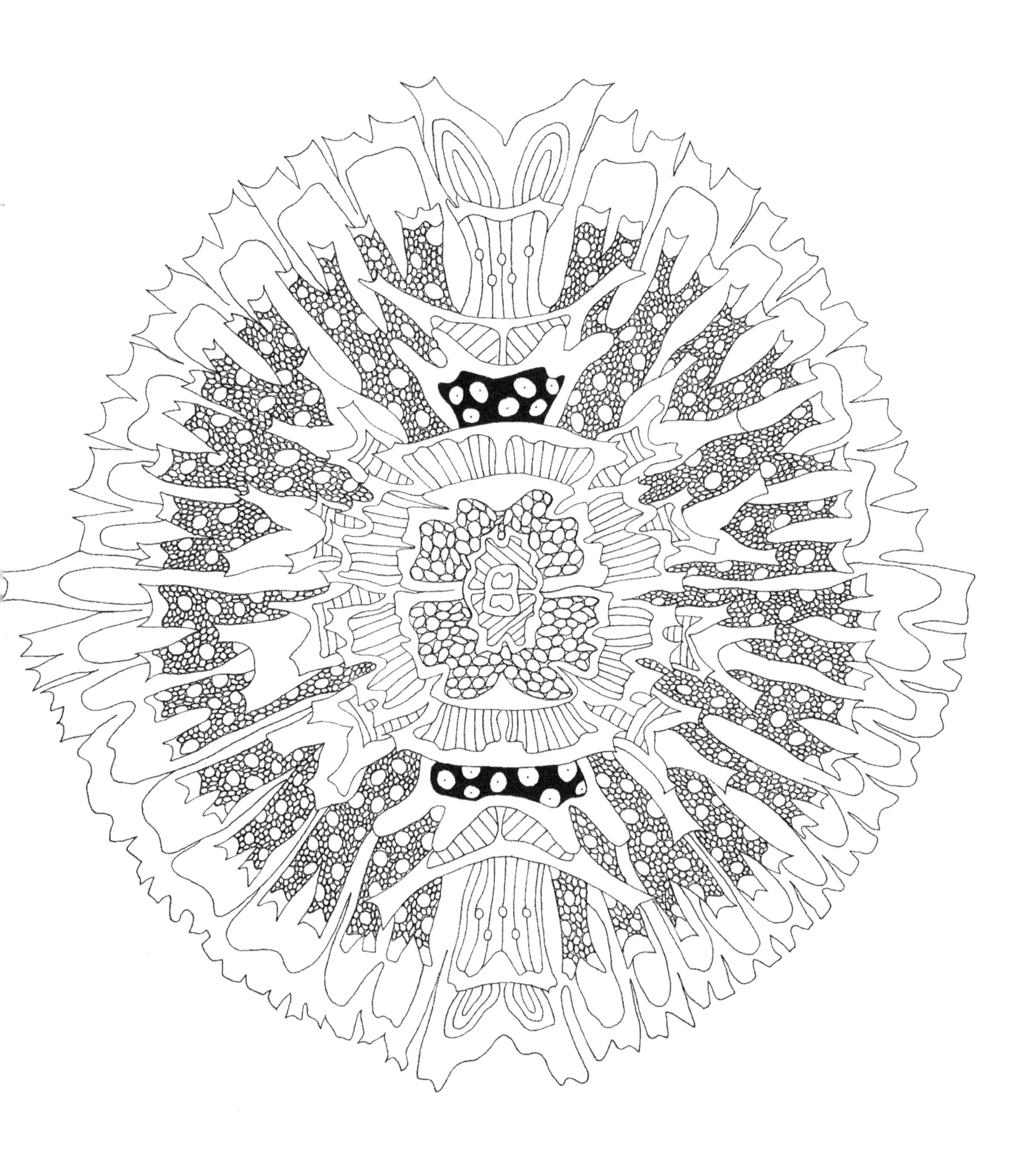

Coccolithophores

Coccolithophores are single celled marine eukaryotic phytoplankton. They are spherical cells enclosed by plates called coccoliths. These are made of calcium carbonate which they pull from the surrounding water. When they die they sink to the sea floor and chalk is formed from their remains.

Coccolithophores live in the sea surface, usually in still, nutrient poor environments. In these nutrient poor areas where other phytoplankton are scarce, coccolithophores provide the main nutrition for many species of zooplankton.

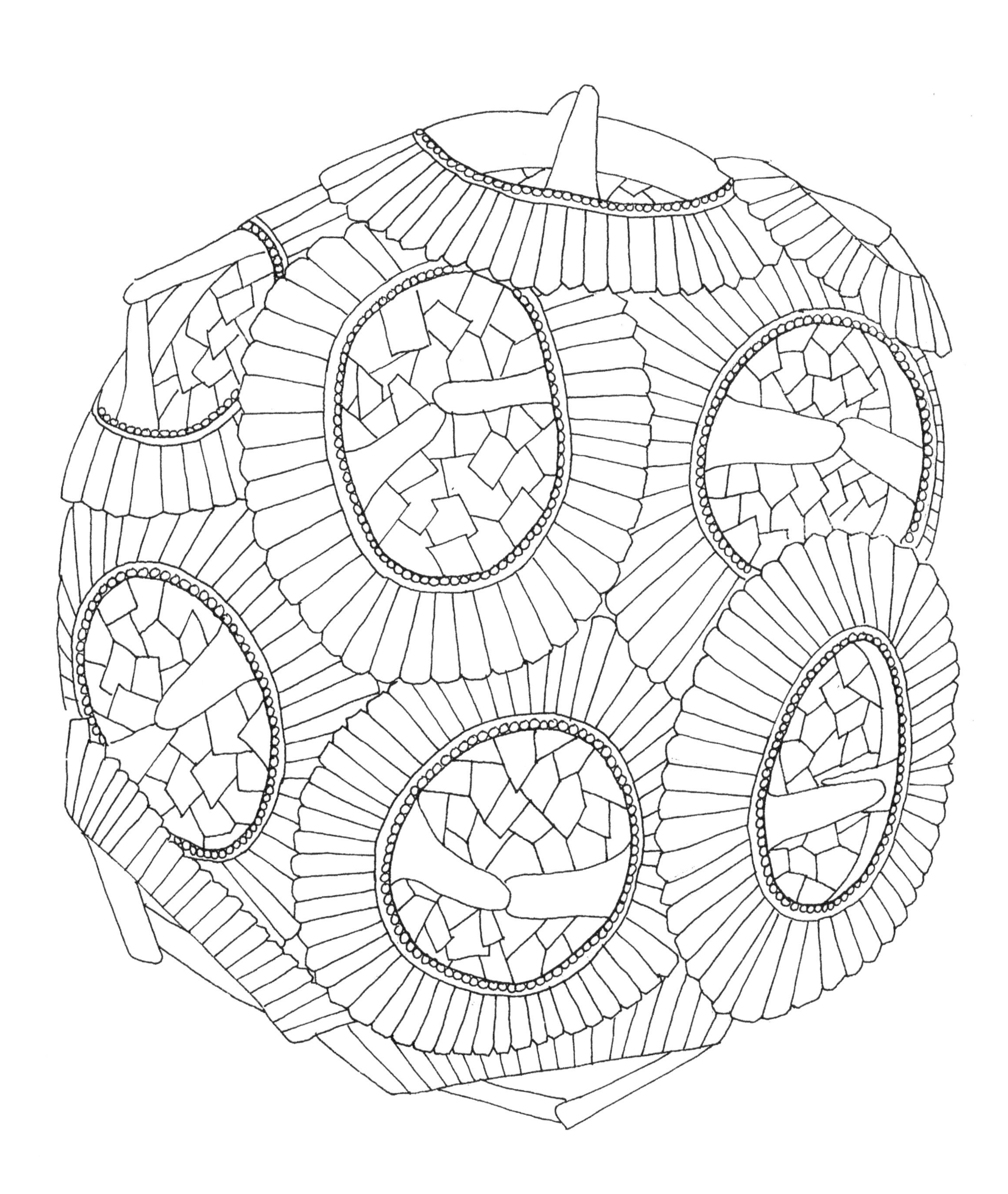

Volvox

Volvox is a freshwater chlorophyte or green algae made up of spherical colonies with up to 50,000 individual cells. These colonies are composed of two differentiated cell types, somatic cells and germ cells.

The numerous somatic cells have flagella which enable them to swim in a coordinated manner. The cells have red eyespots which help them detect light. The colony has a front and a rear with eyespots which are more developed at the front of the colony.

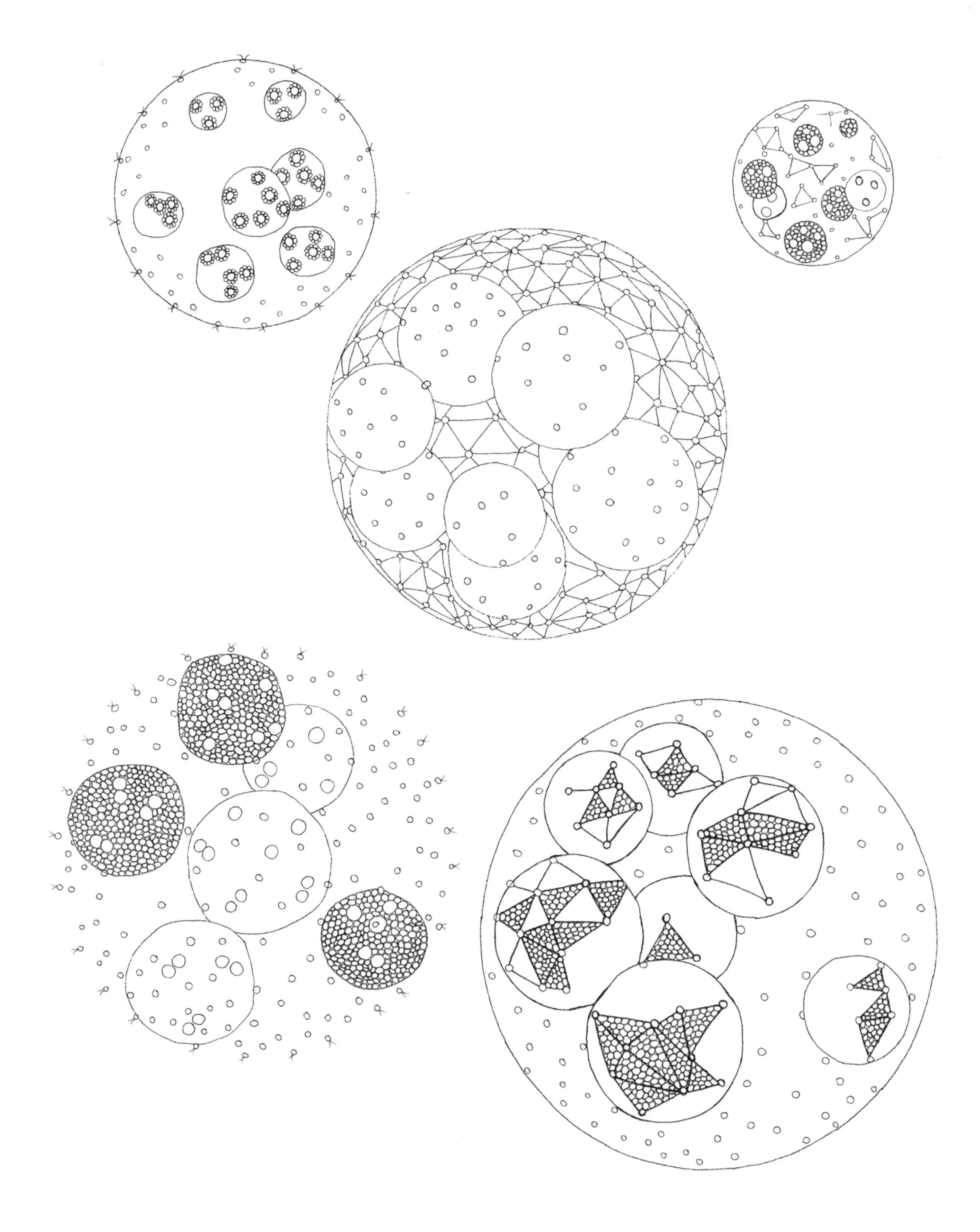

Sphagnum moss

Mosses grow widespread in cool, damp and shaded places. They reproduce via spores and have no flowers. They grow on bare trees and rocks forming cushion shaped clumps.

Mosses have thin leaves that usually spiral around slim, wiry stems. Sphagnum moss, illustrated here, can hold up to twenty times its own weight in water. It lacks stoma, so it cannot control water loss from its tissues.

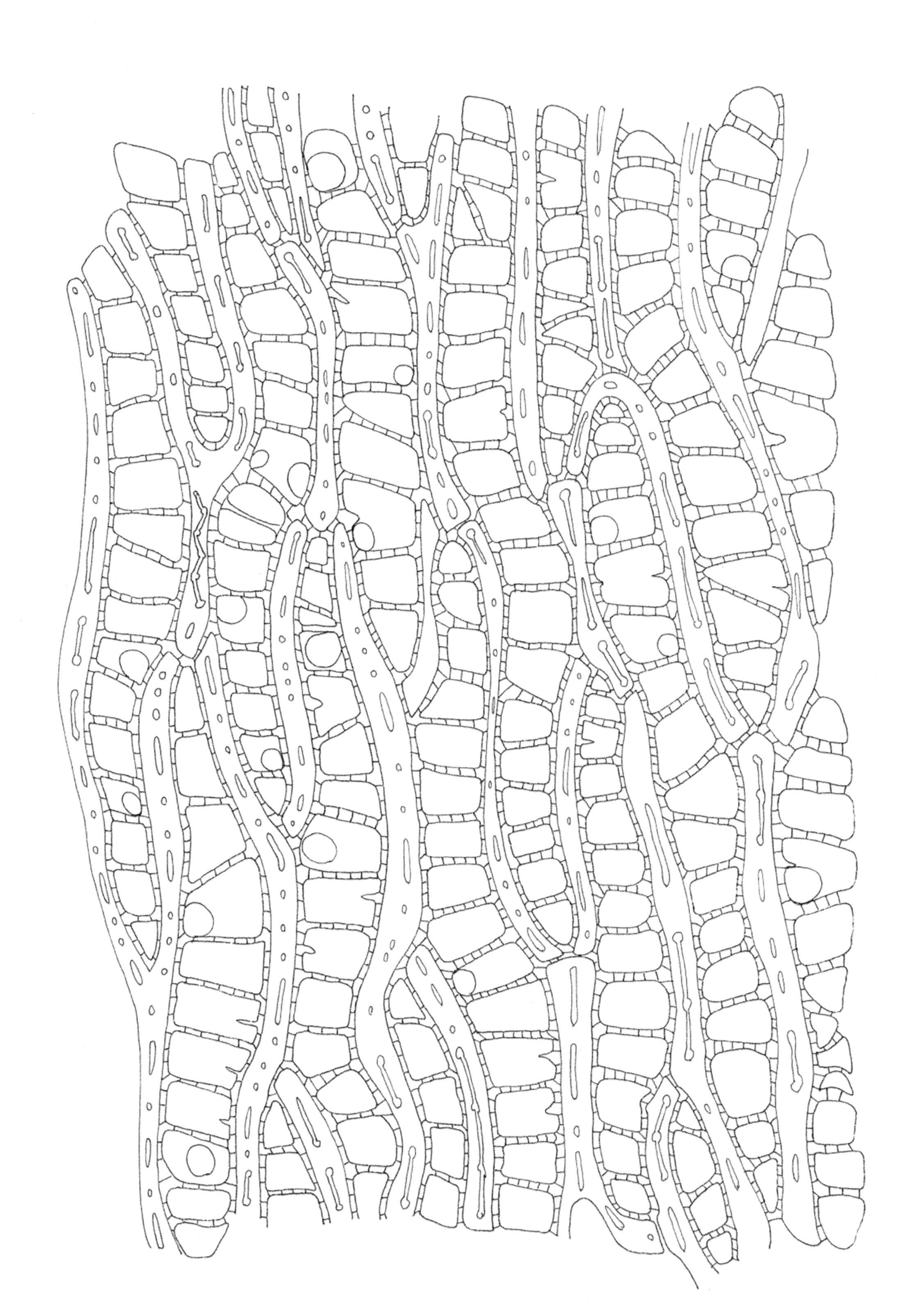

Fern sorus

Ferns are vascular plants that reproduce via spores. They do not have seeds or flowers. Ferns first appear in the fossil record about 360 million years ago. Some species are very small while others, like the tree fern, can grow up to 20 metres in height.

The body of the fern is divided into the rhizome, frond and sporangia.

Spores form on the undersides of the leaves in spore cases called sporangia, as illustrated here. Clusters of sporangia, or sori, appear as brown spots. Some species have sporangia on all their leaves, while others have specialised leaves that bear the sori. When the sporangia dry out, they break open, releasing the spores. Germination begins when a spore falls in a place with favourable conditions.

Ferns have alternating generations of separate spore producing plants and gamete producing plants.

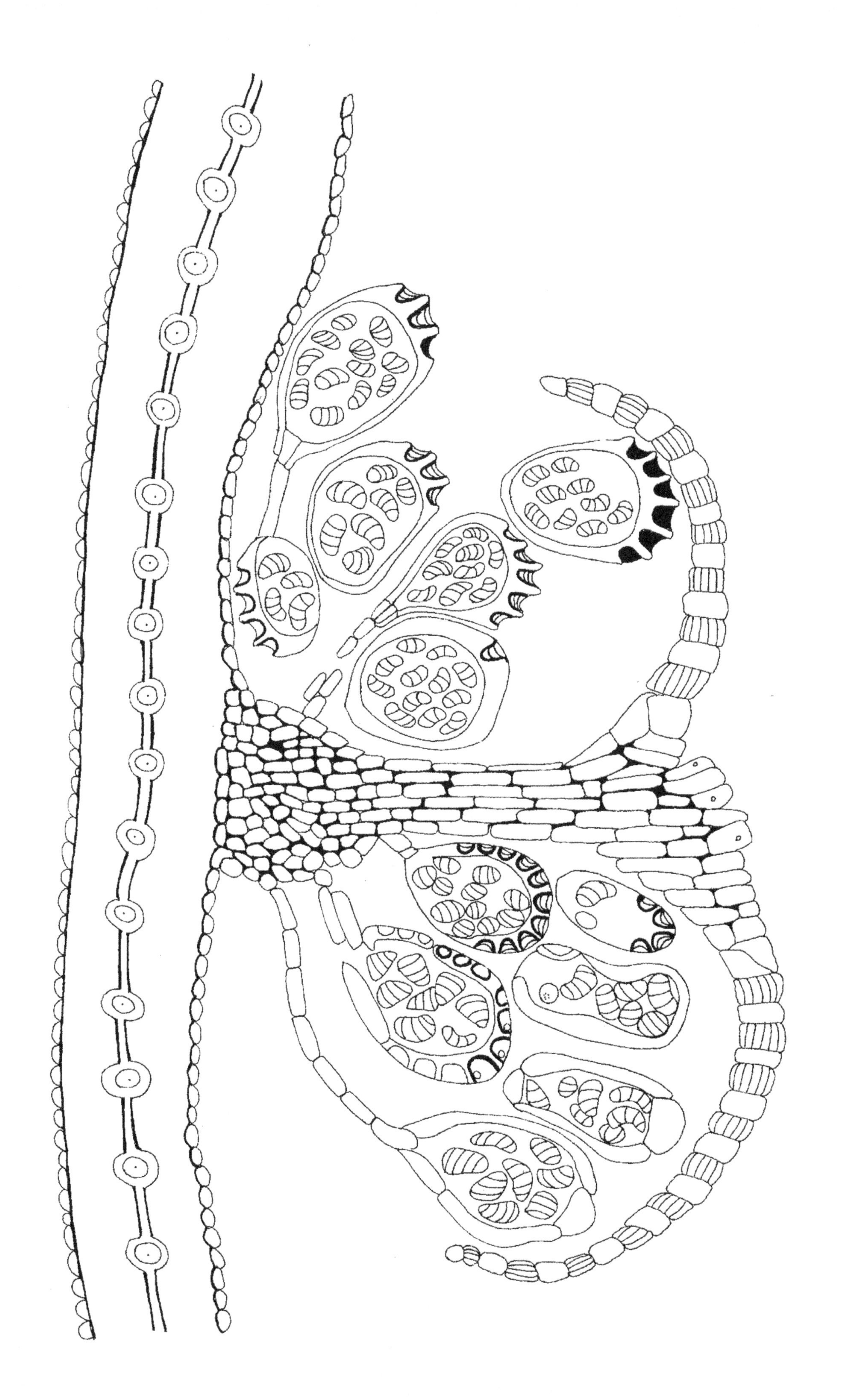

Female pinecone

Pines are the common name of a group of coniferous plants of the family Pinaceae. They are resinous trees which have needle-like evergreen leaves. Native pines are found in most regions of the Northern Hemisphere and many species have been introduced to both temperate and sub-tropical regions around the world.

Pines are gymnosperms, a group of seed producing trees, which have both male and female cones on the same tree. Male cones, which are relatively small and cylindrical, have one or several pollen sacs under each modified leaf called a bract. Their pollen is released when conditions are right and transported by wind to the female cone.

Unlike male cones, the female pine cones are more conspicuous and are retained on the tree for longer periods. Their overlapping scales are arranged in Fibonacci ratio spirals. Each scale has two ovules. The scales are covered in bracts which give them their brown familiar appearance.

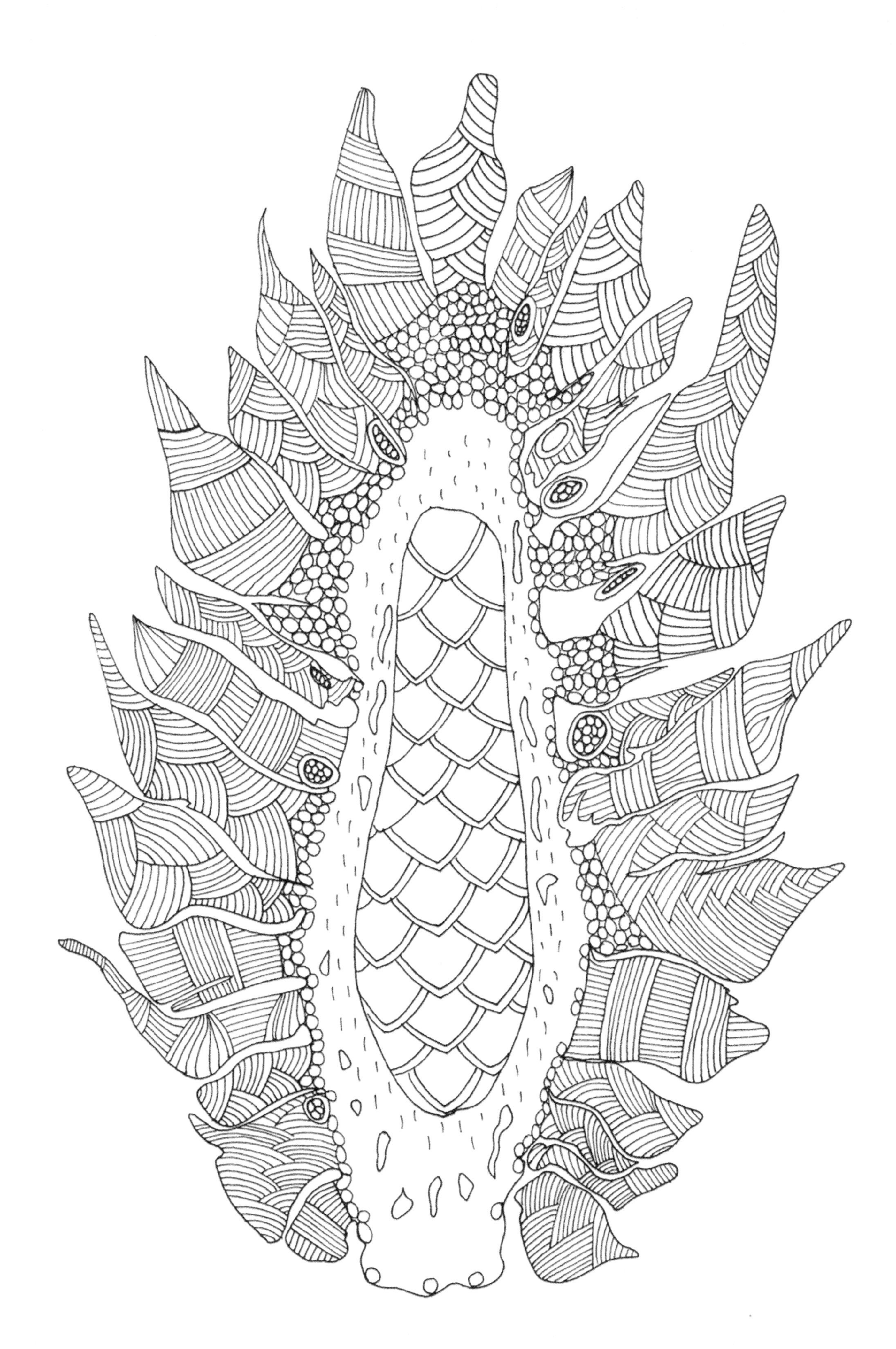

Male pinecone

In spring or early summer, the male cone releases pollen from the male pollen sacs. Wind carries the pollen to the female pinecone which opens temporarily to receive the pollen. Fertilisation occurs the spring of the following year. When seeds are developing, female cones are kept closed tight and sealed by a resin. The seeds mature within six to eight months in a large, dark and woody pinecone which opens up to release the seeds when conditions are favourable.

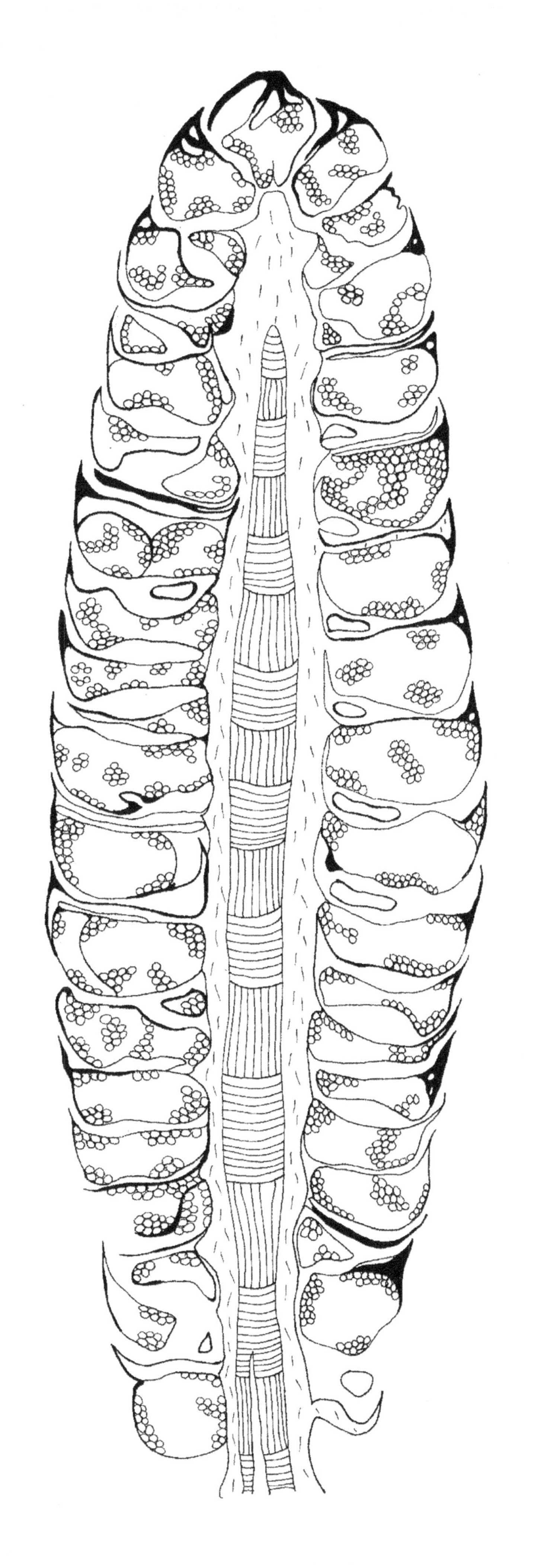

Lily cross section

Lilies are well known herbaceous flowering plants which have prominent, often beautiful flowers. They are perennial flowers growing from starchy rhizomes, corms or bulbs. There are 250 genera and 4000 species of *Lilium* which show great diversity in colour, especially in the numerous cultivated hybrids. They are native to temperate regions of the Northern Hemisphere, though some are found in the tropics.

Lilies are monocots, which means that their sepals, petals, anthers and ovaries are arranged in threes. The ovary is the female reproductive organ of the lily flower. The ovary is part of the pistil that holds the ovules, and it connects with the base of the petals and sepals. One single ovary is split into three chambers called locules. It is here that the ovules are contained.

The cross section of the lily shows three outer sepals and three inner petals. Six stamen house the anthers while the central carpels house the ovary and ovules.

Lily anther

The anthers of the lily plant release the male gametes in the form of pollen. Pollen lands on the style and stigma which is above the ovary. The pollen germinates and grows through the style to the ovary. It takes only one pollen grain to fertilise one ovule.

Once fertilised, the ovary wall develops into a fruit. Fruits can be fleshy, hard, multiple or single. Seeds germinate and the embryo grows into the next generation.

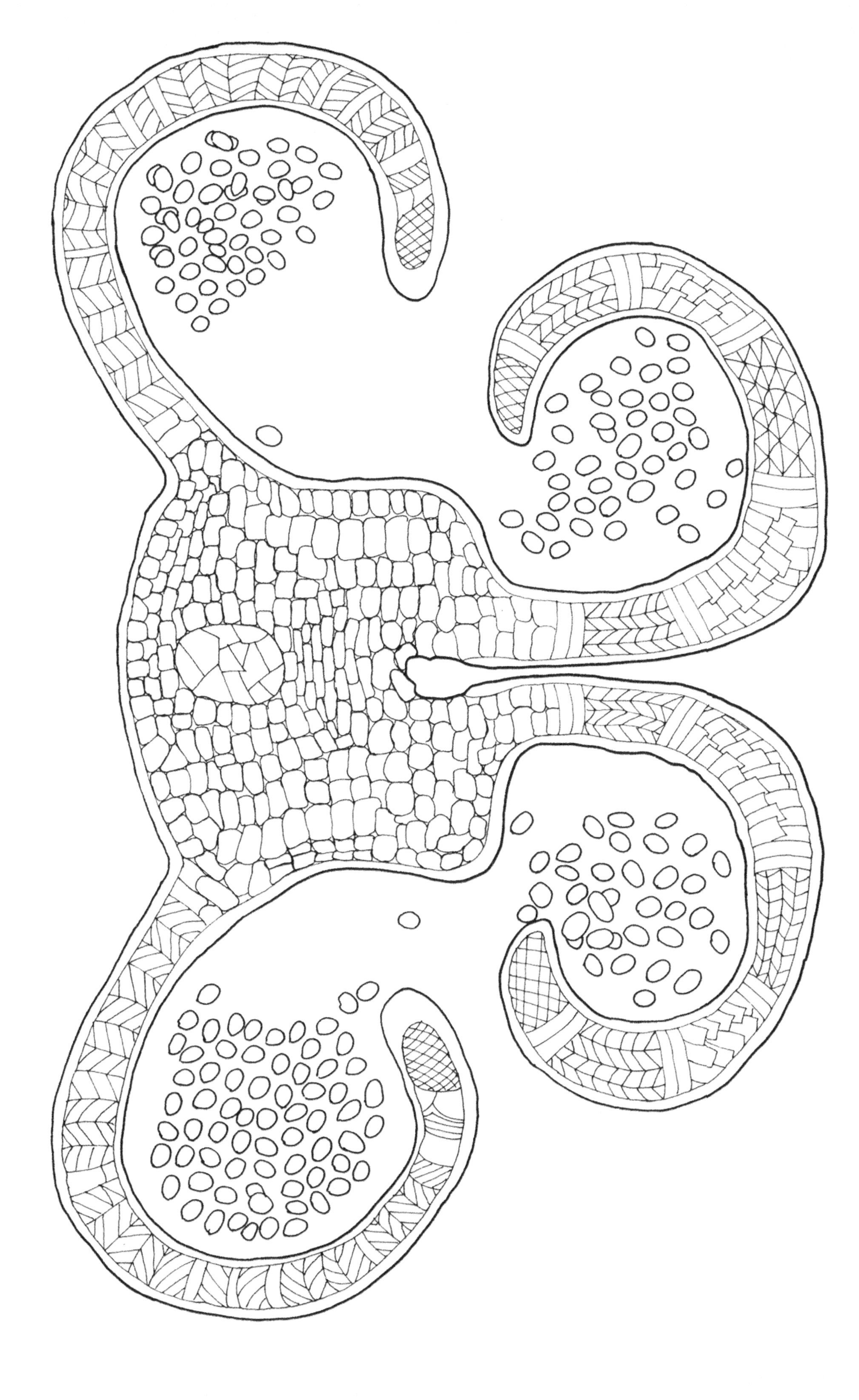

Passiflora cross section

Passiflora is a genera of flowering plant, mostly in the form of vines which are found in tropical regions of continents including South America, New Guinea and Asia. Many species have been used for their medicinal properties.

This cross section of *Passiflora caerulea* shows five outer petals, five stamens holding the anthers and a central ovary.

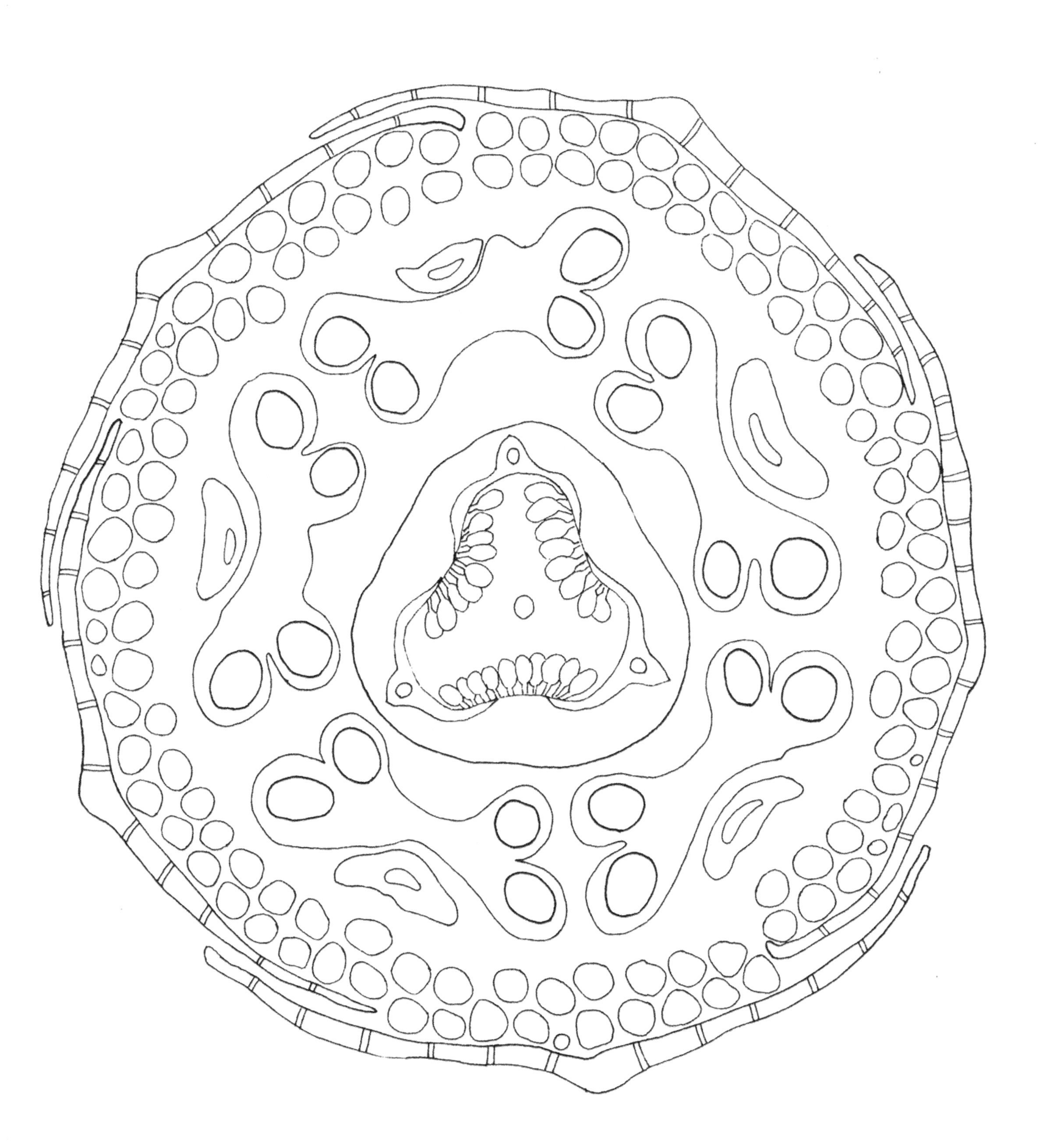

Tilia cordata.

Tilia is a small leaved lime tree which is native to most of Europe. It is an ornamental tree with distinctive heart shaped leaves. This illustration shows a cross section of a stem of *Tilia cordata.* Notice the different structures within the stem. The outer bark protects the inner phloem, cambium, xylem and pith.

The phloem transports food from the leaves to the rest of the plant. The xylem transports water and solutes from the roots of the plant to the leaves. The vascular cambium which lies between these layers is responsible for the secondary growth of stems and roots which creates width to the plant as it grows.

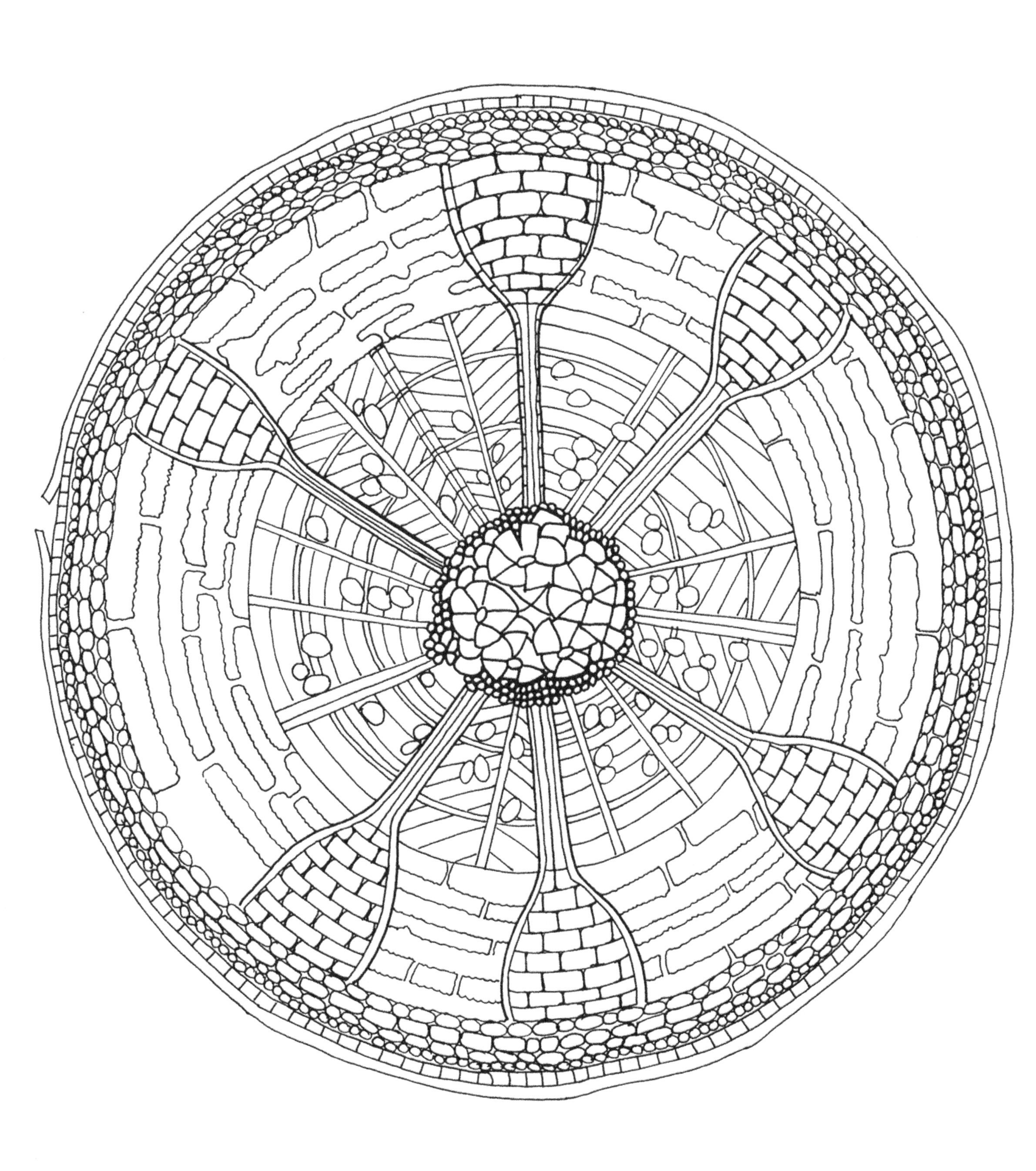

Milkweed cross section

Asclepias, commonly known as milkweed is an American genus of herbaceous perennial. There are 140 known species with some of the most complex flowers in the plant kingdom. Milkweed are a source of nectar for native bees and wasps in America.

This illustration shows a central ovary containing two carpels. The stamens, filaments and styles are all fused together and surrounded by five hoods and five outer petals.

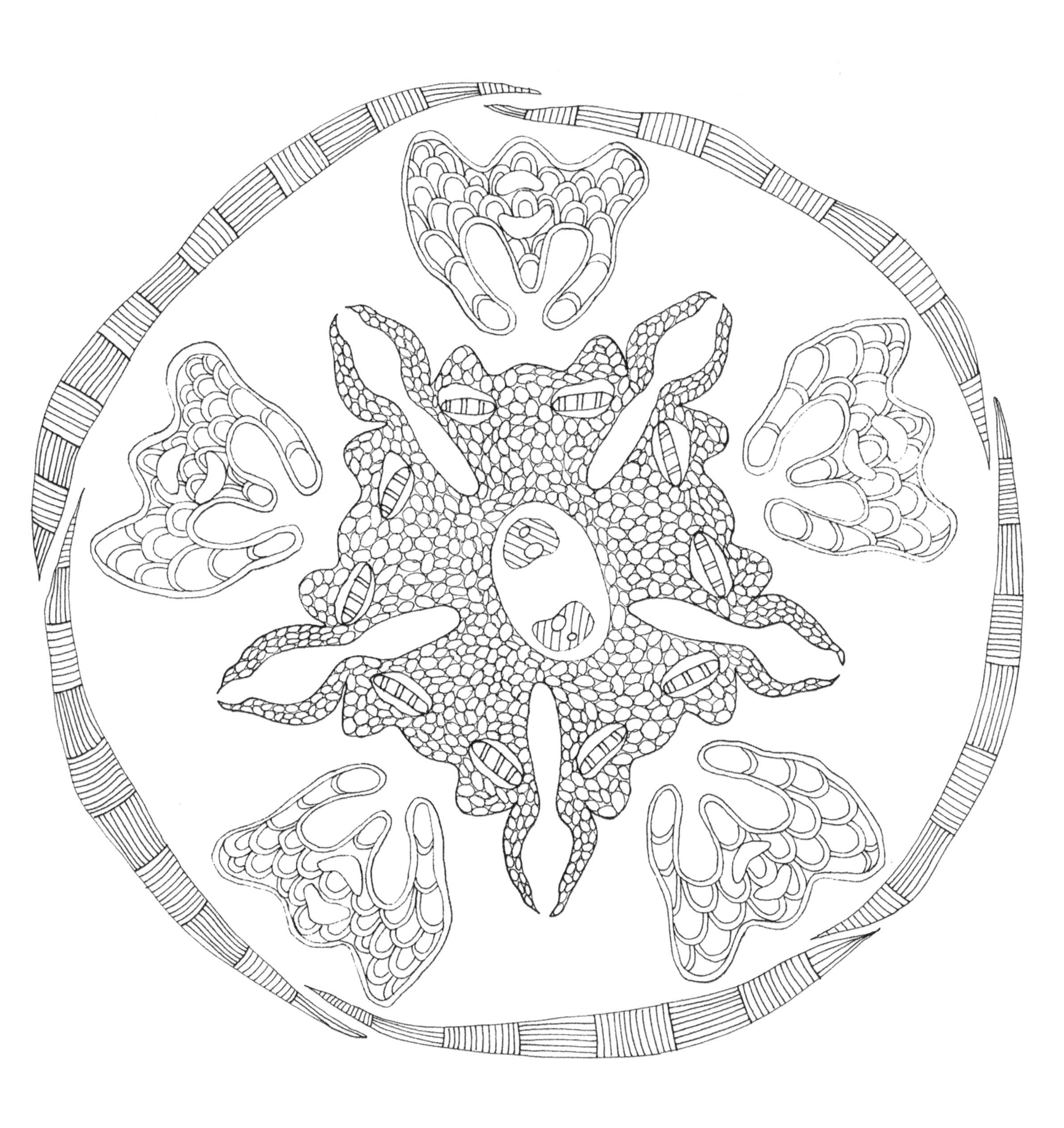

Passiflora pollen

Pollen grains come in a wide variety of shapes and sizes. While they are often spherical, it is their size and surface ornamentation that distinguishes between different species. Some pollen grains can even be winged. The study of pollen is called palynology, and it is useful in forensics, archaeology and paleontology.

Pollen grains have a double wall. The stamen of flowering plants releases pollen to germinate the pistil, or in conifers the male cone releases pollen to fertilise the female cone. If the pollen lands on a compatible pistil or female cone then it can germinate.

Pollen has a hard coat that protects the male gamete from drying out and from solar radiation. Pollen is transferred from one plant to another by cross pollination or from the anther to a stigma of the same flowering plant, via self-fertilisation.

The pollen of *Passiflora setacea* illustrated here, shows a spherical grain with three distinct ring-like apertures. The ornamentation is described as reticulate.

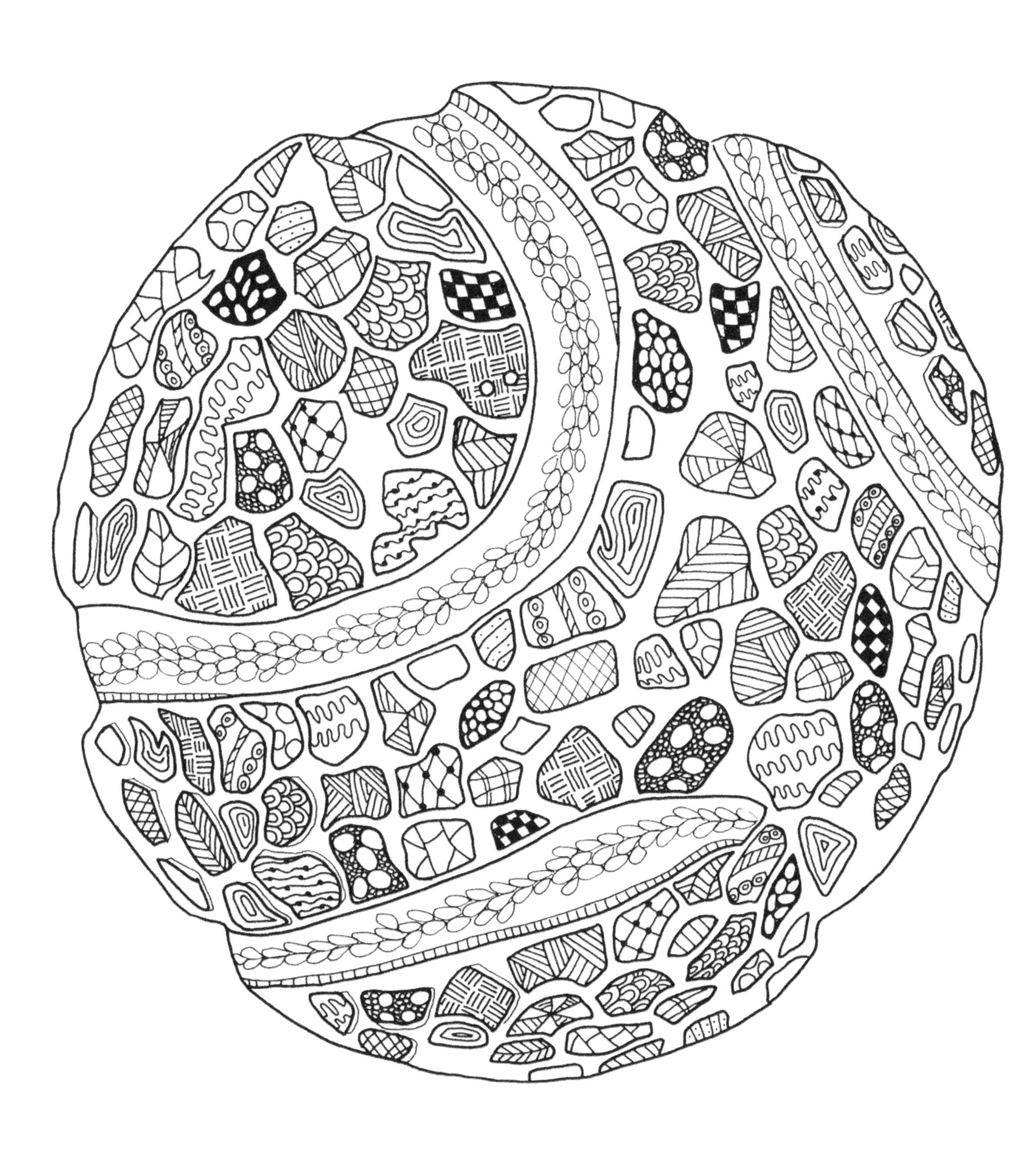

Acacia pollen

There are about 800 different species of *Acacia*, found mainly in warm tropical areas, with many species native to Australia. They belong to the pea family and have clusters or spikes of yellow or white flowers. Their wood is used in furniture and their flowers are used in the food and perfume industries. Their pollen can form distinctive patterns in some species.

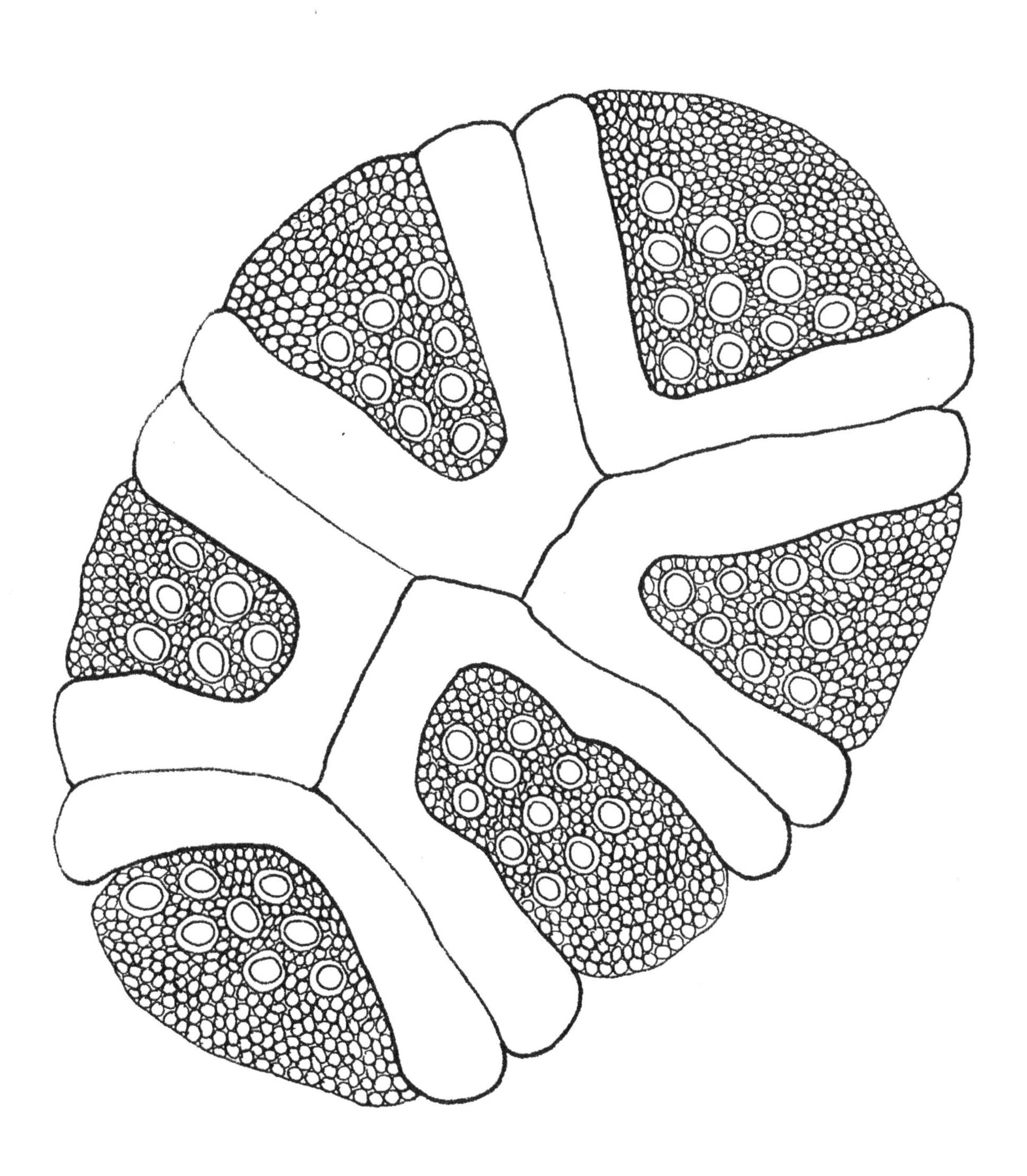

Trichodesma africanum seed

Trichodesma africanum is a flower found in tropical and sub-tropical areas of Africa, Australia and Asia. These flowers from the borage family have medicinal properties. They can be used to treat hepatitis and inflammatory conditions.

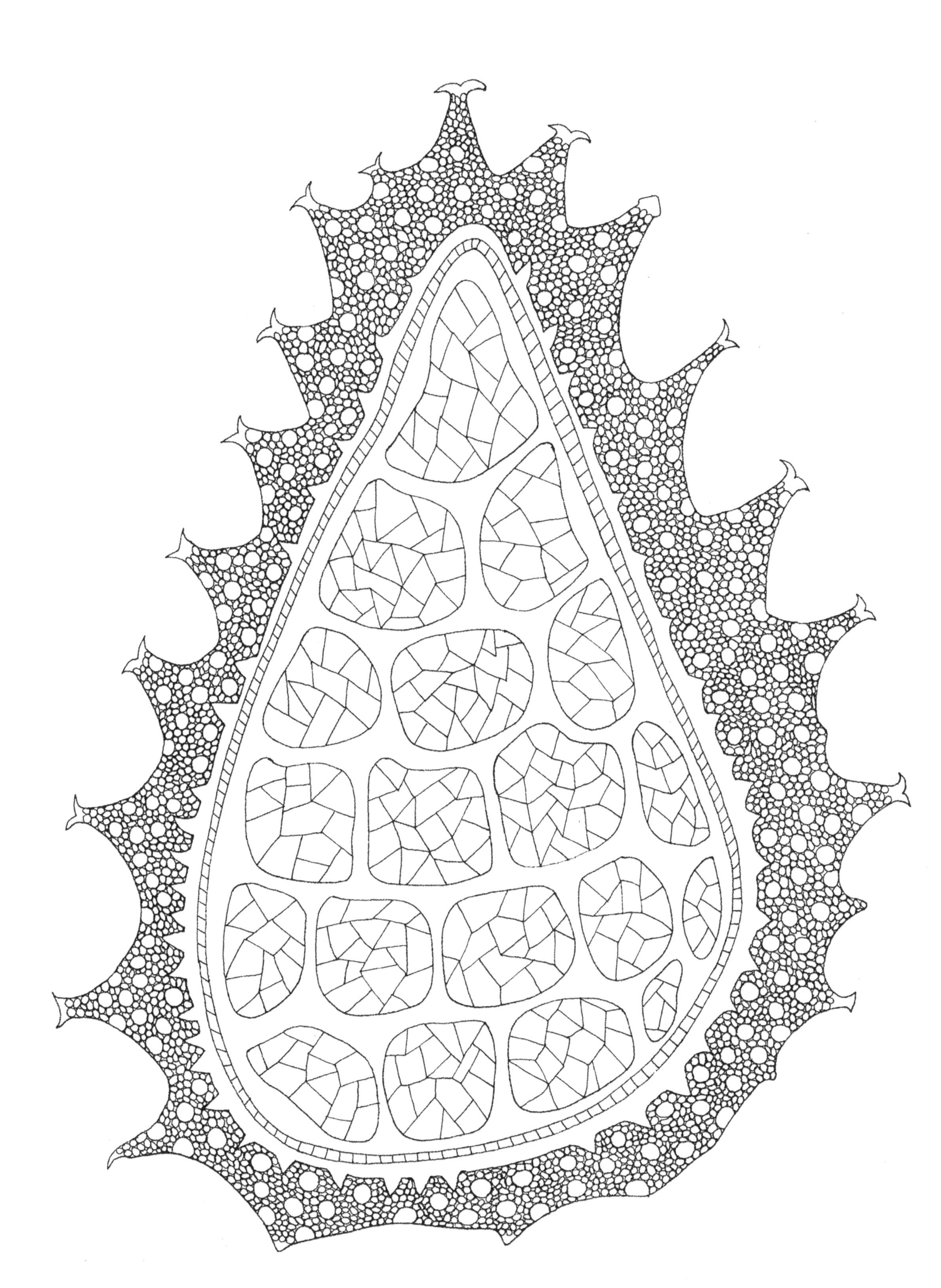

Scorpiurus muricatus seed pod

Scorpriurus muricatus is an angiosperm and is part of the legume family. It is commonly found as ground cover in Southern Europe and some parts of the Middle East. It has small pea-like flowers and simple leaves but its seed pods are contorted given rise to the common name of "prickly caterpillar".

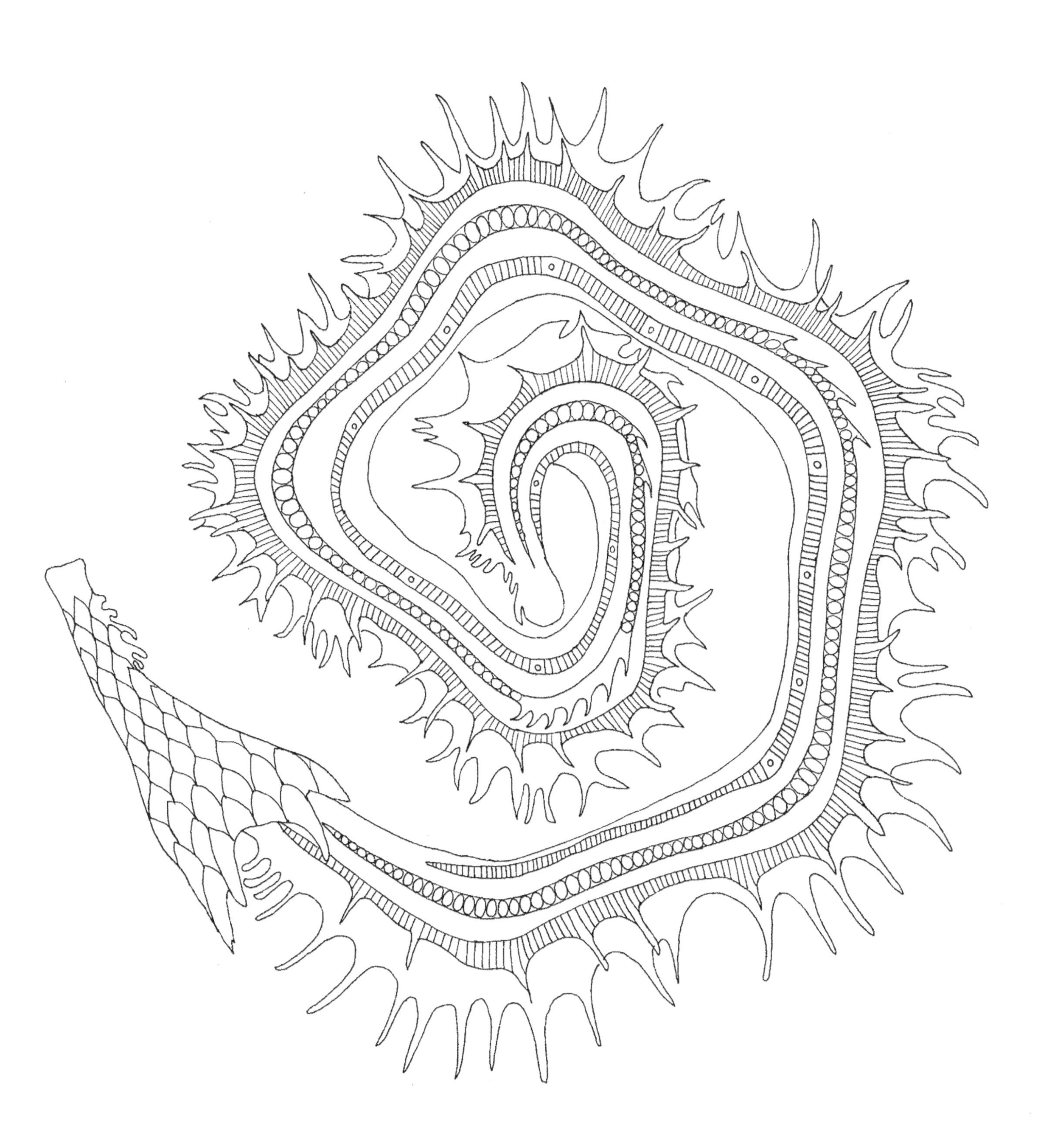

Cup fungus

Peziza is a cup fungus that grows on the ground, on rotting wood and on dung. The vertical section of the apothecium, illustrated here, is where sexual reproduction takes place. The apothecium is made up of mycelium with a basal hypothecium. The latter consists of fertile asci containing ascospores and sterile paraphyses.

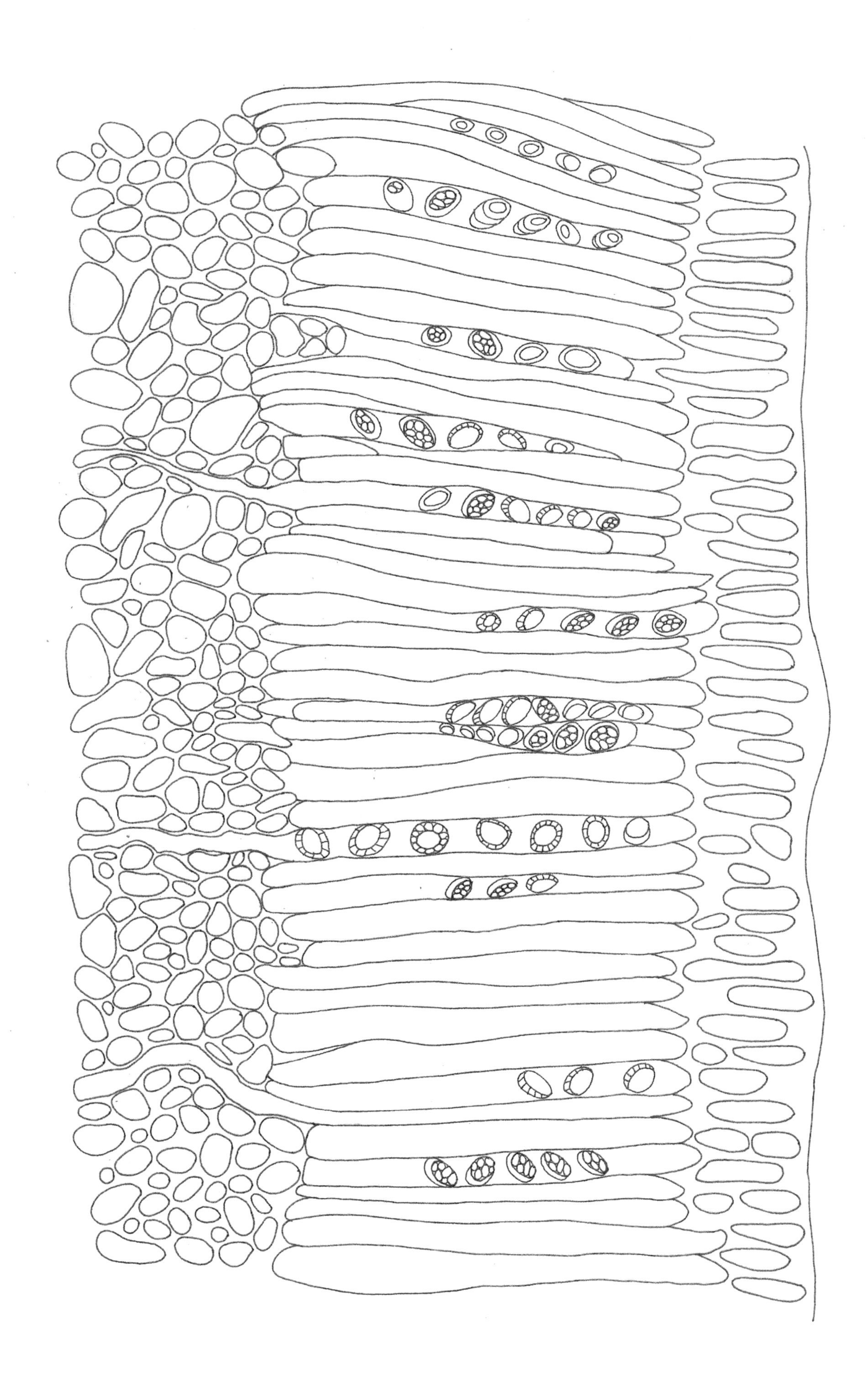

Aspergillus flavus

Aspergillus flavus is a conidial fungus found in oxygen rich environments. It is a contaminant of cereal grains, tree nuts and legumes. It can cause aspergillosis in people with a weakened immune system, those suffering from allergies or those with damaged lungs.

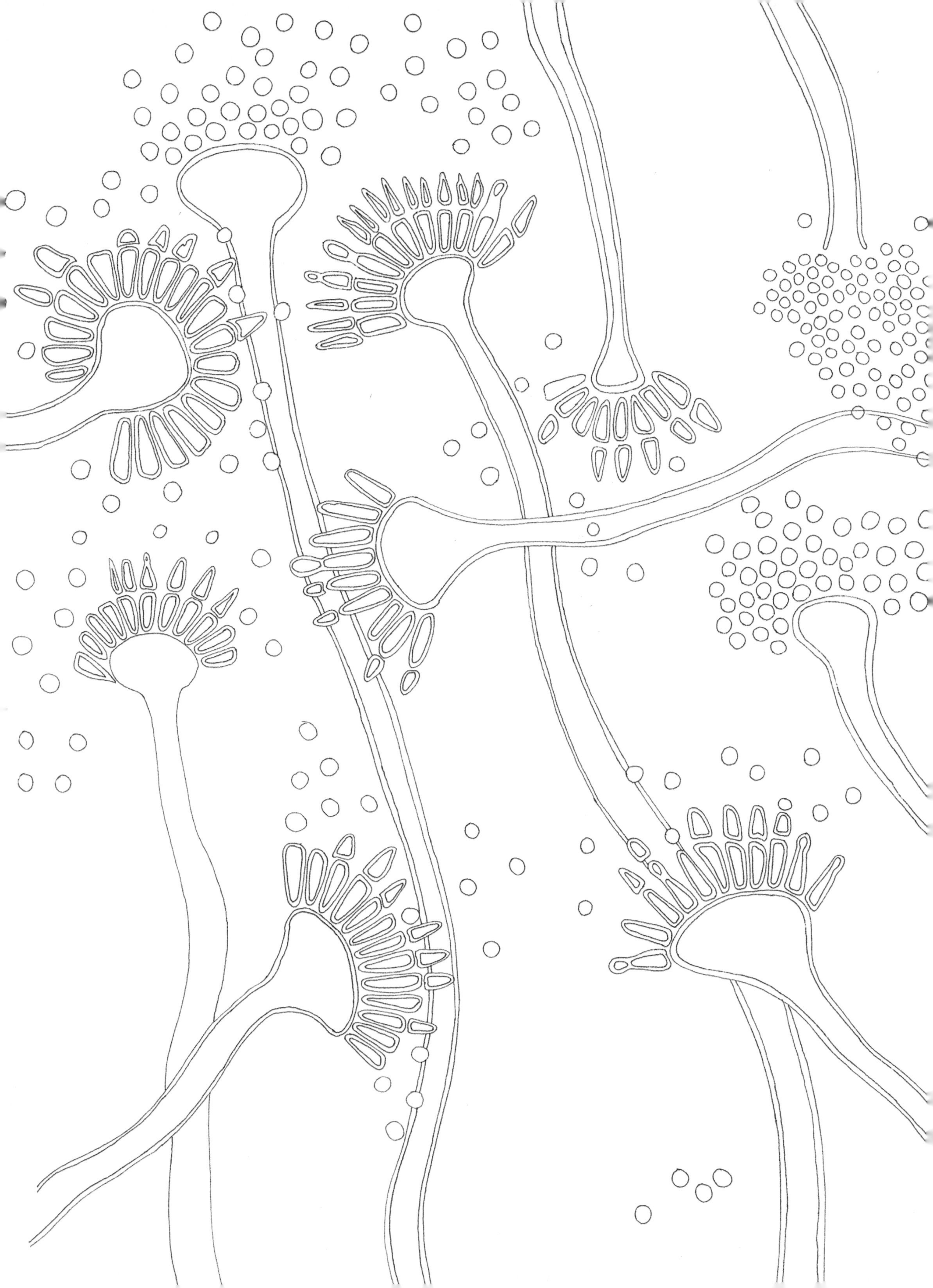

Moon jellyfish ephyrae

Aurelia aurita can be found in the Arctic, Atlantic and Pacific oceans. It is a translucent jellyfish, usually about 25-40 cm in diameter with four horse-shoe shaped gonads which make it easily recognisable.

It has an asexual benthic stage in its life cycle called a polyp which produces ephyrae. These ephyrae, illustrated here, develop into either a male or a female adult medusa. In time, the male medusa release sperm which fertilise the eggs from the female medusa and form a free-swimming planula larva. It eventually settles on the sea floor and develops into a polyp and the cycle continues.

Platyhelminthes

Platyhelminthes, also known as flatworms, are simple invertebrates consisting of three main cell layers and exhibit bilateral symmetry. Their flat shape means that they respire via diffusion and they possess no body cavity, therefore the same opening takes in food and expels waste.

Many flatworms have quite complex life cycles, requiring more than one host to complete their life cycle. Over half of all know species of Platyhelminthes are parasitic. They can inflict vast harm to humans and to livestock.

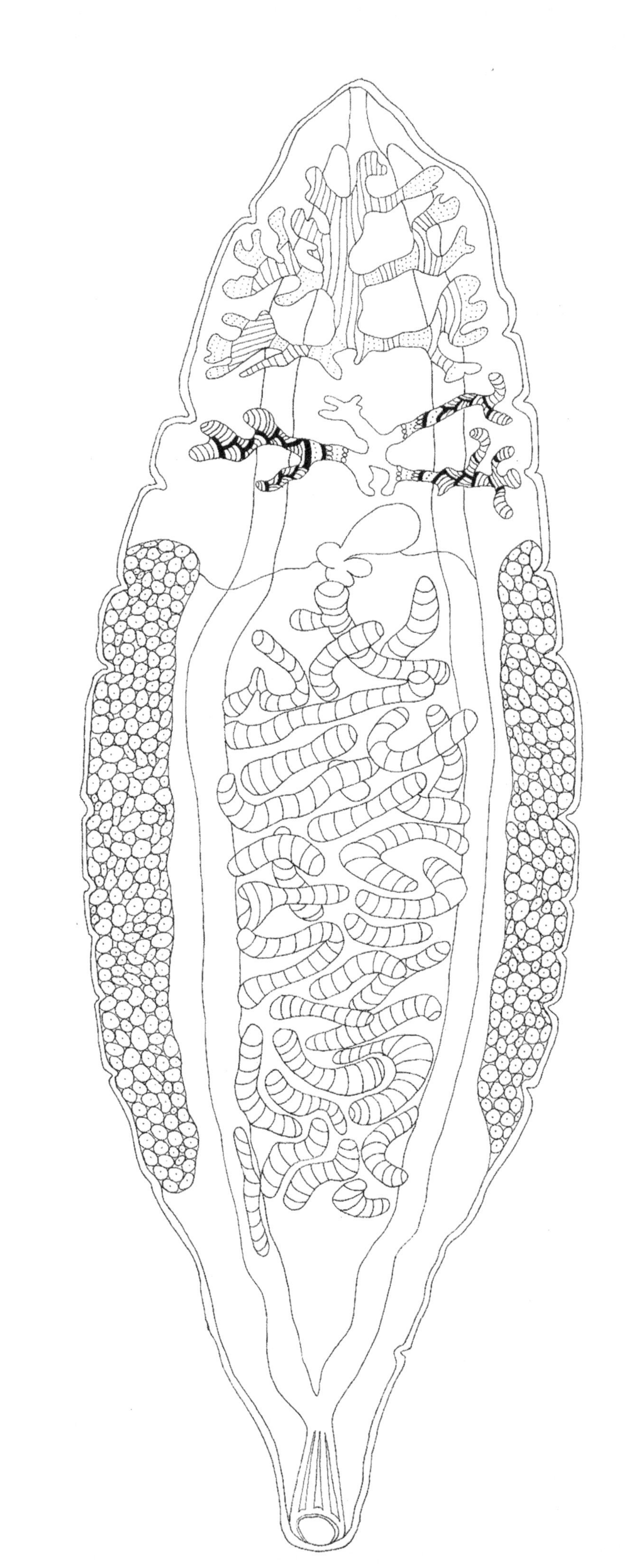

Rotifers

Rotifers are microscopic multicellular animals found in freshwater and moist soils. They are commonly found on mosses and lichen. Their bodies consist of a head, neck, trunk and foot.

The mouth possesses a crown of cilia which draws in water which they sift for food. Rotifers consume unicellular algae and phytoplankton as well as dead and decaying organic matter. Food is ground down by jaws called trophi, located in the pharynx.

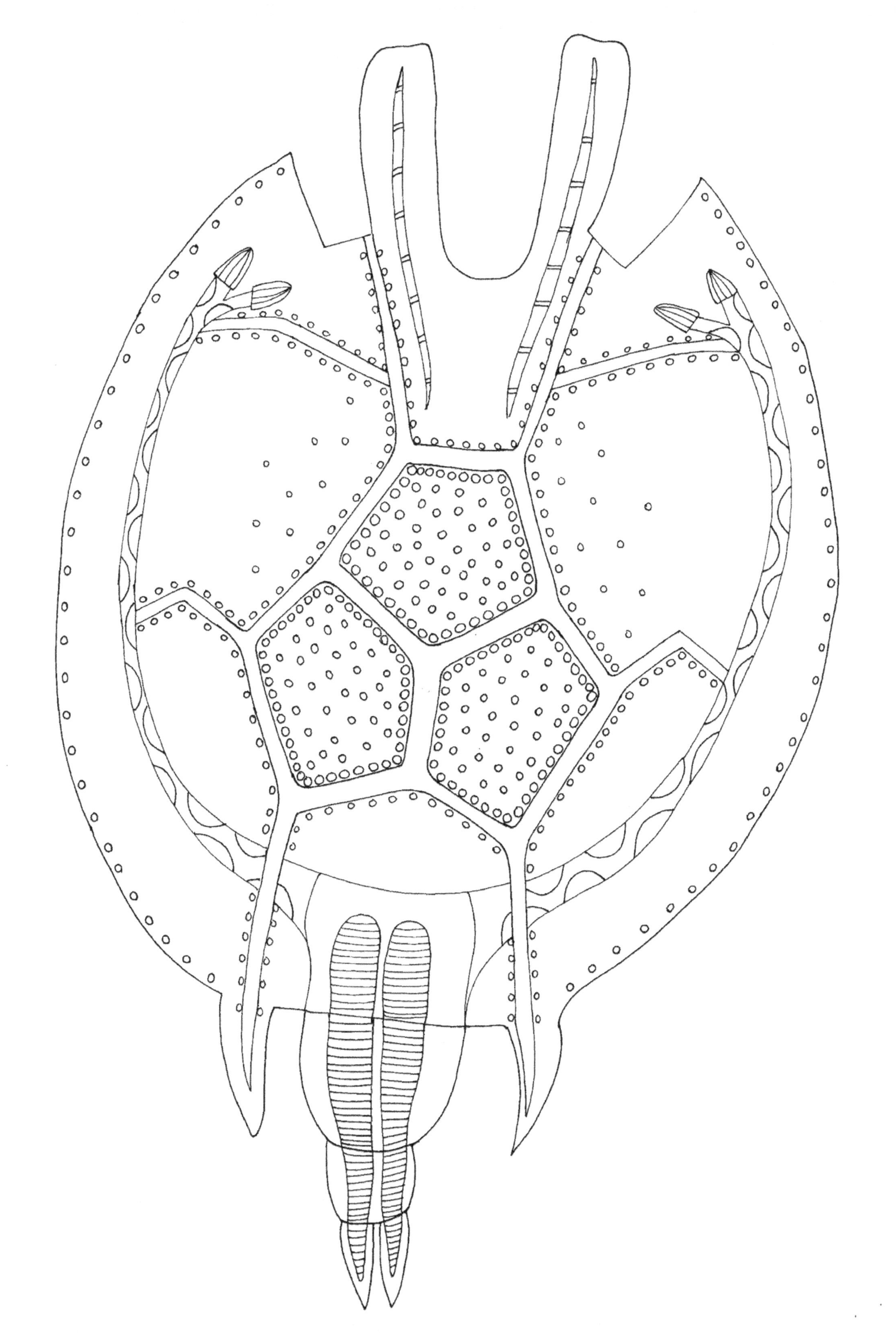

Loricifera

Loricifera are very small to microscopic marine metazoans that live in spaces between marine gravels. They have only been discovered relatively recently and many species have yet to be described. Their bodies are divided into five distinct regions, a mouth cone, introvert, neck, thorax and abdomen. The introvert containing the mouth cone can be retracted into its abdomen. The abdomen is covered in thick lorica which consists of plates which provide protection.

Loricifera are the first multicellular animals to live their complete life cycle in the absence of light and oxygen. They have been found 3000m deep in the Mediterranean Sea.

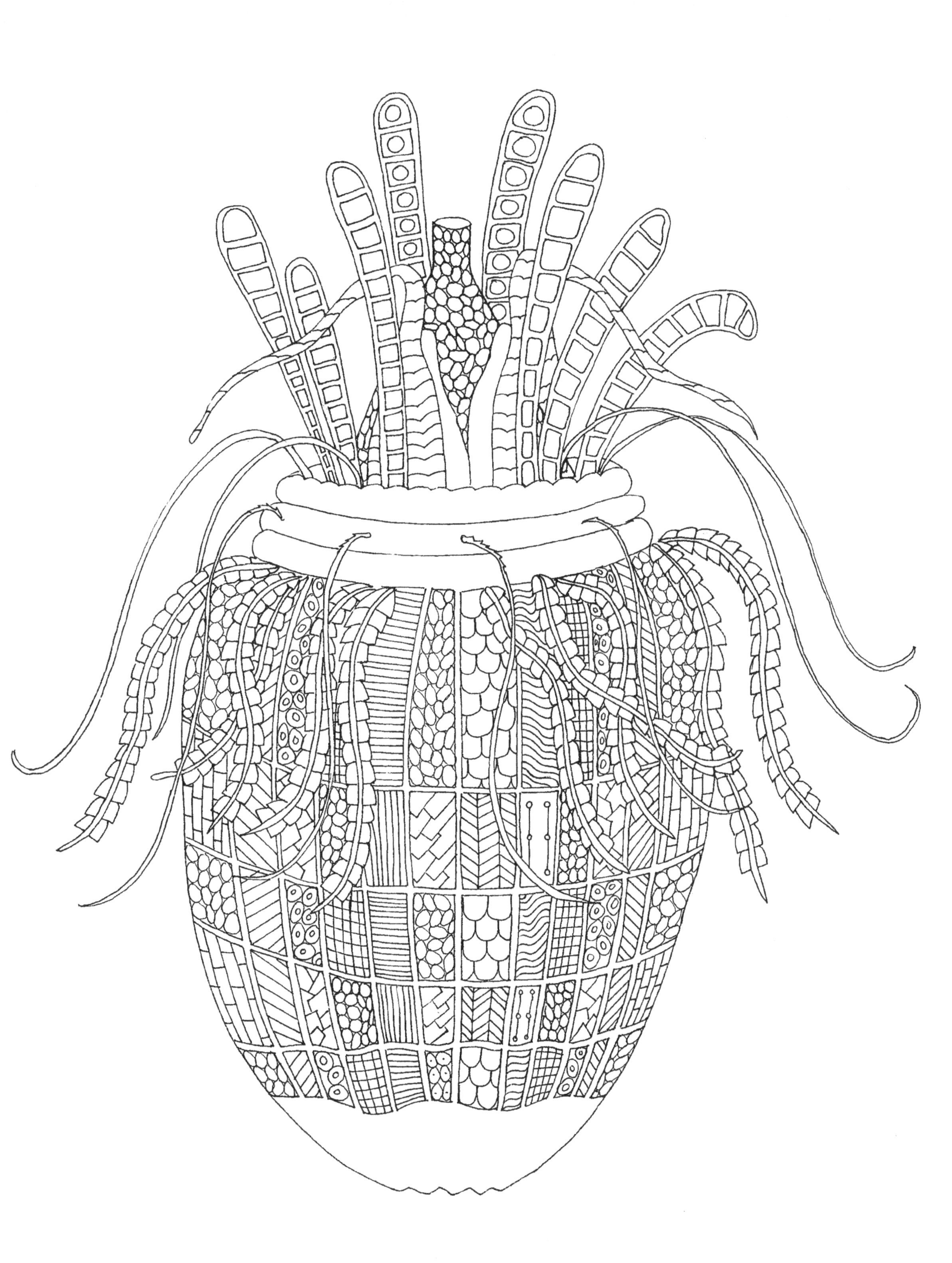

Tardigrades

Tardigrades, commonly known as water bears, are microscopic, water-dwelling, segmented animals known as one of the most resilient on Earth. They can be found on the top of mountains and deep in the ocean floor. Tardigrades can withstand extremes in temperature from -272ºC to 150 ºC as well as radiation hundreds of times that which man can tolerate. They can also withstand extremes of pressure, from the vacuum of space to six times the pressure of the deep sea.

Water bears have a barrel shaped body divided into a head and three body segments, each having a pair of legs. The legs bear claws called disks. Their barrel shaped mouths have stylets used to bite plant cells, other small invertebrates and algae. They have been known to survive without food or water for thirty years.

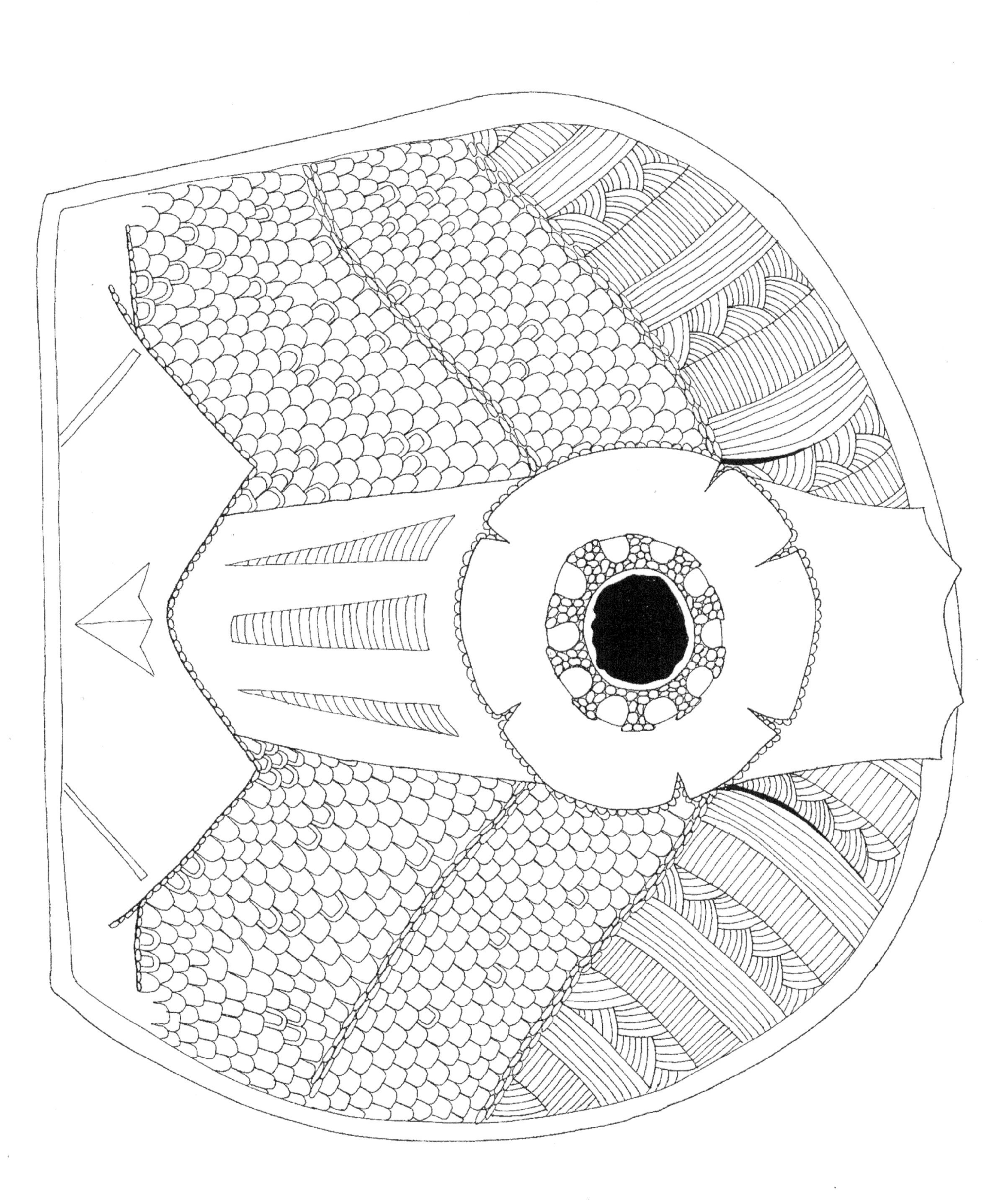

Rocky mountain dog tick

Dermacentor andersoni, also known as the Rocky Mountain dog tick, is a dangerous vector of Rocky mountain spotted fever, Colorado tick fever virus and the bacteria that causes hunter's disease. They are resilient animals that can survive 600 days without feeding. Their short mouthparts are characterised by a central hyposome which is lined each side by palps and cutting chelicera.

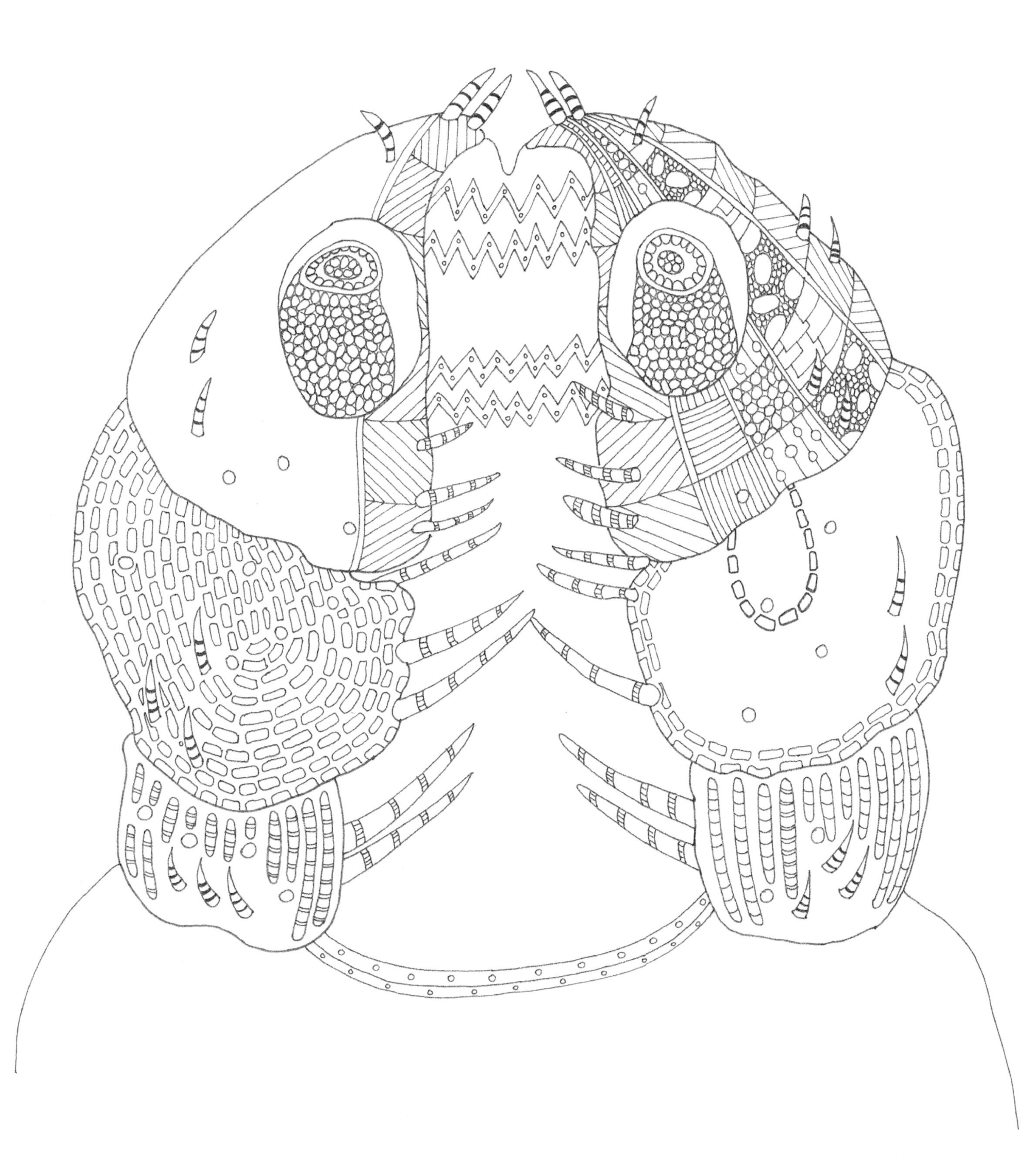

Daphnia pulex

Daphnia pulex is the most common species of the group of organisms known as freshwater fleas. They are free living aquatic animals which live in freshwater habitats. They live as plankton in open water of lakes or can be attached to vegetation. Life in open waters would prove too difficult for these delicate crustaceans.

The body of *Daphnia pulex* is not clearly segmented but it has a noticeable head and body. The head has a dark compound eye and two pairs of antennae which are used in feeding and swimming. The mouth has a crushing mandible that grinds filtered plankton such as small algae, bacteria and fine detritus. Most species of *Daphnia* are herbivorous but some species are carnivorous and prey on other water fleas.

The body has a transparent folded carapace that covers the organs. Food can be easily seen passing through the intestine of the animal to its posterior opening. The central part of the body is called a thorax, having five pairs of flattened legs which are covered in setae (hairs). The outer carapace ends in a spine. Males are generally smaller than females but have longer antennules.

Daphnia pulex live usually for 10-30 days but can live up to 100 days if there is a lack of predators. They are preyed upon by fish such as sticklebacks, minnows and young sockeye salmon. They are important in the freshwater food chain as they convert algae and bacteria into a more usable form of food.

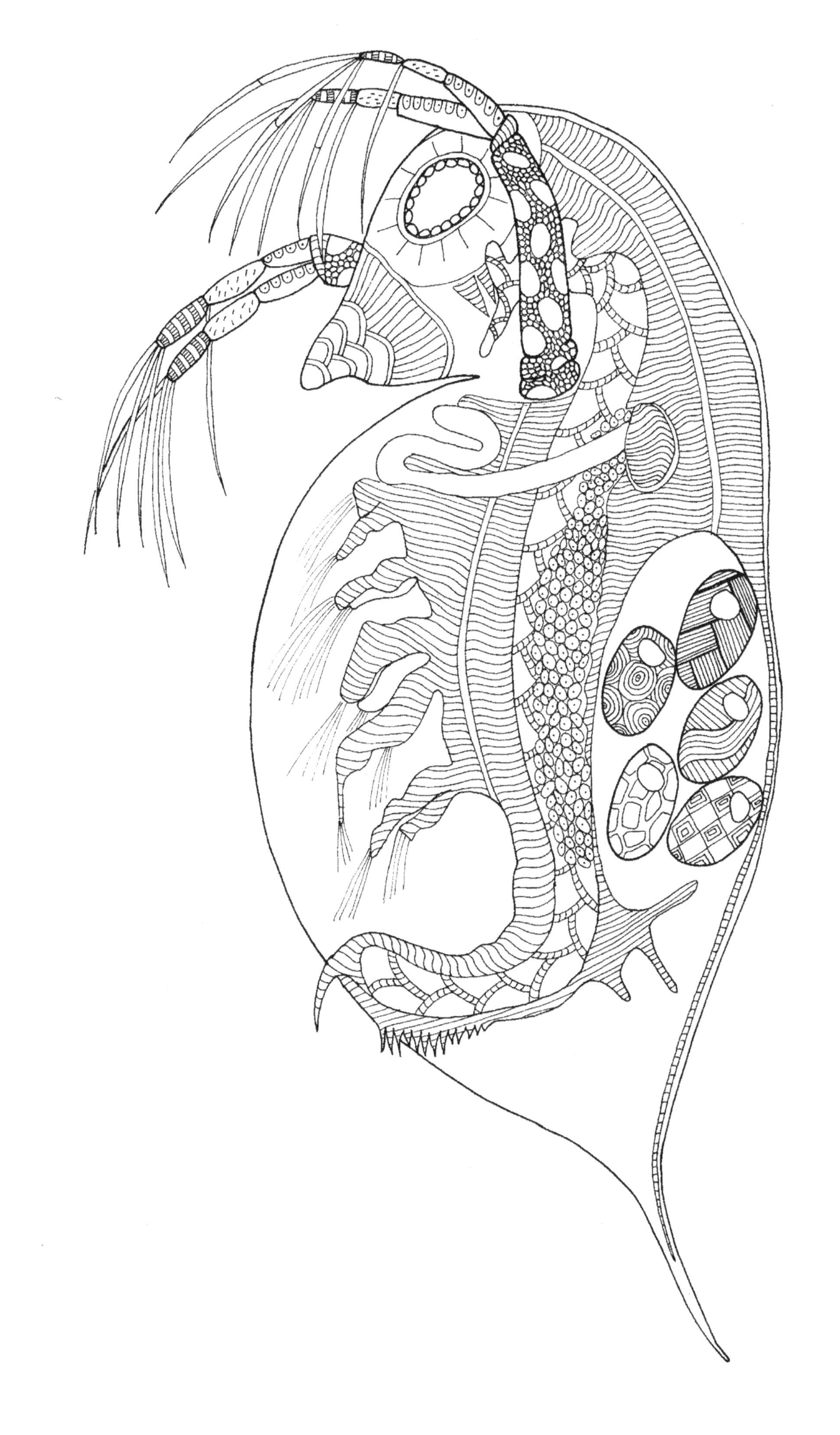

Barnacles

Barnacles are exclusively marine animals which usually live between 5-10 years in shallow and tidal waters. They cement themselves head-down to rocks, boats, whales and crabs.

They are sessile suspension feeders who filter plankton and detritus from their surroundings using six pairs of thoracic limbs called cirri. These cirri, illustrated here, are derived from six pairs of swimming legs in the cypris larva. Each cirri bears two branches, with each segment having bristles, or setae which overlap when feeding. The three longer cirri stroke the surrounding water repetitively to gather food which is scrapped off by the three shorter anterior pairs of cirri. The food is then passed into the mouth which is deeper within the shell.

Barnacles compete with limpets for space. Their main predators are whelks. Certain types of barnacle are eaten in Japan, Spain and Portugal.

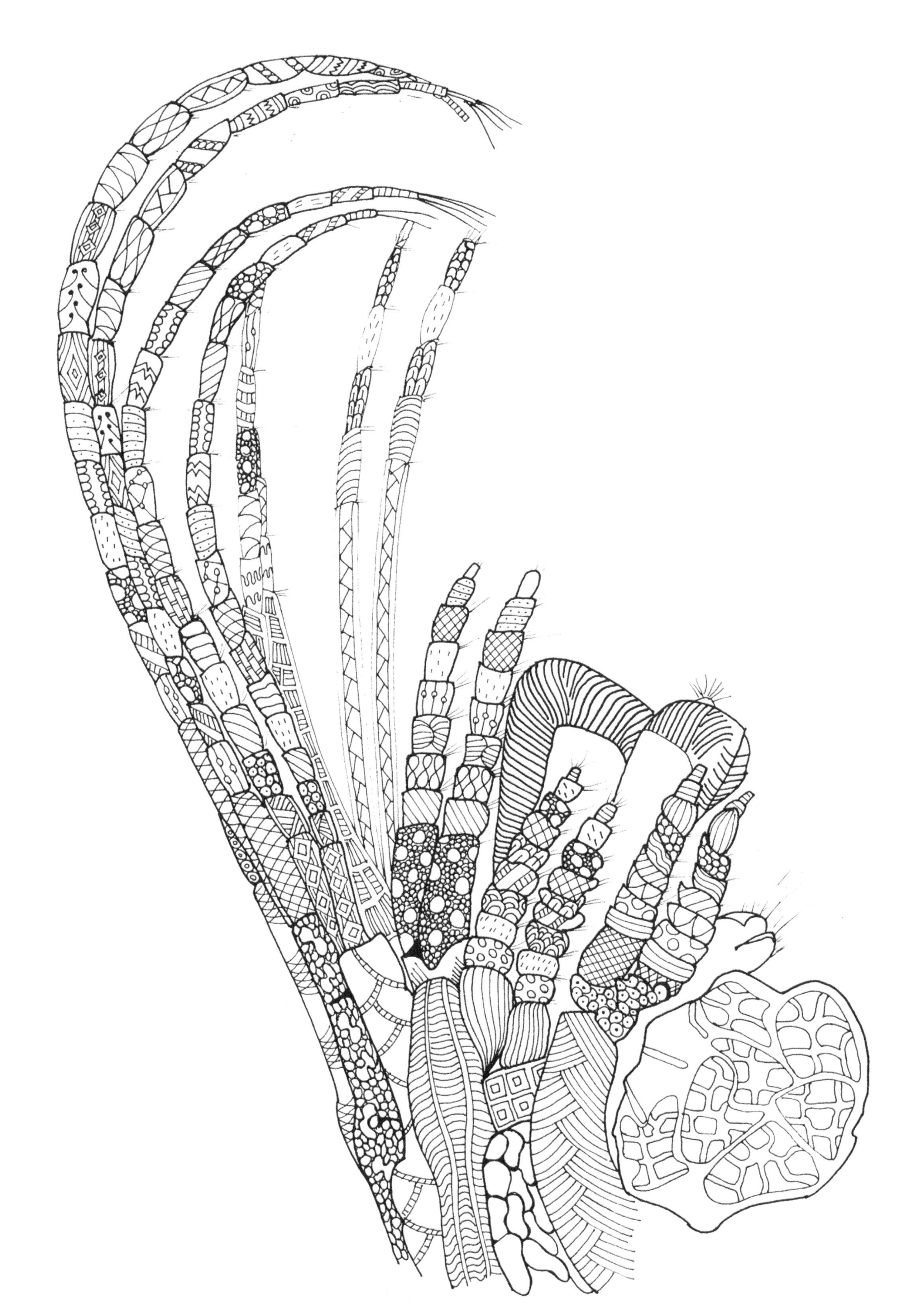

Barnacle larva

Barnacles are hermaphrodite but self-fertilisation is rare. Reproduction is usually by cross fertilisation which occurs between adjacent individuals. Once fertilised the barnacle egg develops into a Nauplius larva, as illustrated here. Nauplius larvae are characteristic of many crustaceans. These one eyed larvae swim using their setae, or bristles. They live to about six months before molting into a cyprid, which is the next larval stage. The cyprid explores its surroundings with modified antennules to find a suitable place to settle. It tests water chemistry and relative moisture before attaching itself using its antennules. They grow directly onto the substrate and finally metamorphose into young adults.

Crab zoea

Crabs are invertebrates that belong to the decapod family. There are over 10,000 species of crab living in our oceans, in fresh water and on land. Crabs possess an exoskeleton made of chiton which doesn't allow for body growth so it is shed through molting.

Reproduction in crabs occurs when eggs are fertilised on the female and these eggs undergo a series of molts before reaching their adult form. After each molt, their size and structure changes, so much so that they were once thought to be different species.

The first larval stage is called a zoea, which is illustrated here. These possess a large cephalothorax, which is a fusion of the head and thorax. It is covered in a helmet shaped carapace often decorated with spines. A noticeable beak-like projection called a rostrum protrudes downwards at the front. Its head is adorned with two large compound eyes. It has short antennules and antennae which are sensory in function. The first and second appendages on the cephalothorax are used for swimming and feeding while the thoracic appendages are non-functional. The abdomen consists of six segments, with a forked extension at the end of the telson (tail).

Zoea larvae are common during the summer and autumn. They capture food material suspended in the water column and are themselves preyed upon by other suspension feeders.

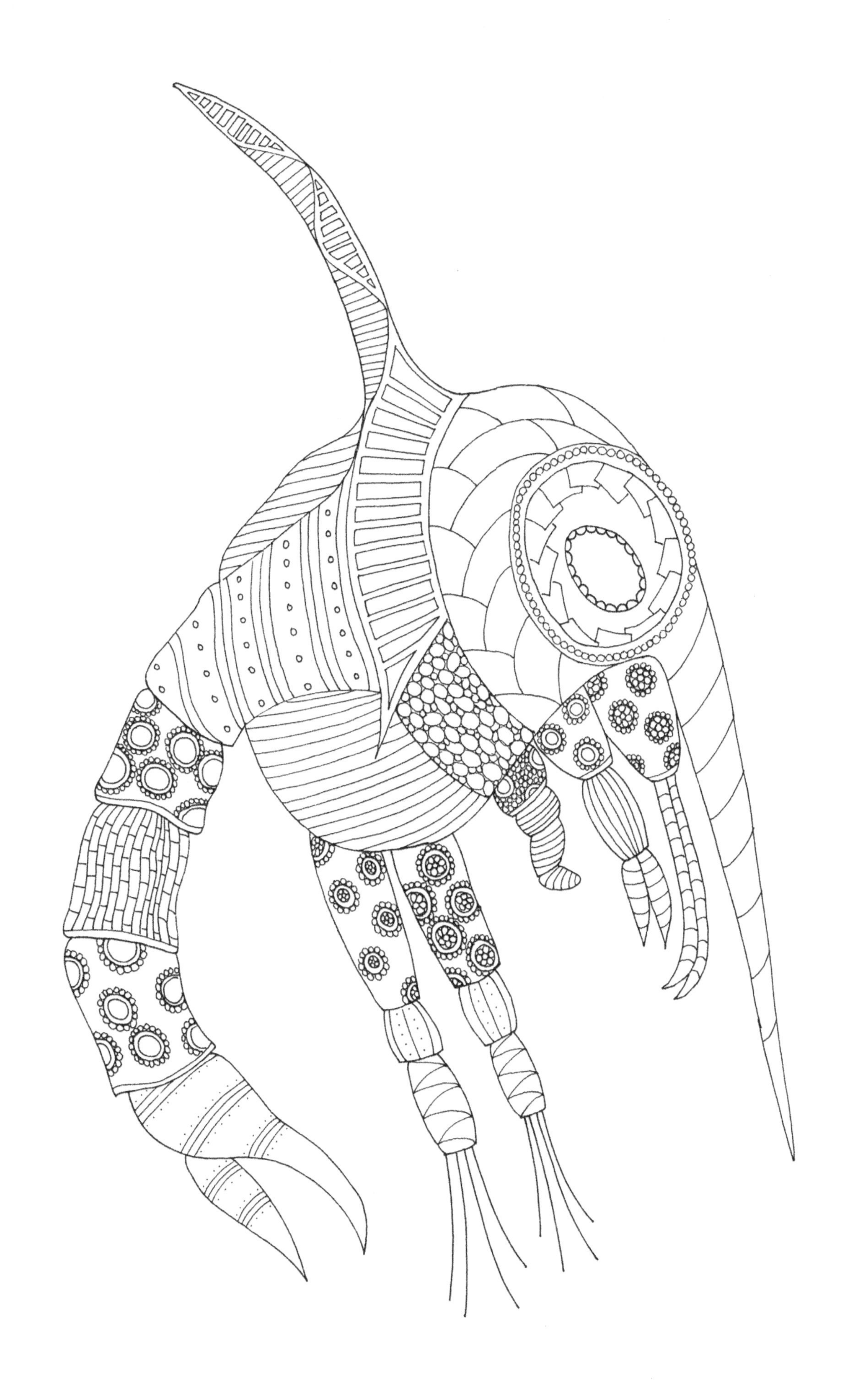

Copepods

Copepods are a small group of crustaceans found in both marine and freshwater environments. They exist in both planktonic and benthic forms in the world's oceans as well as in wet terrestrial places, caves and sinkholes.

Copepods are typically 1-2 mm long. They have a tear drop shaped body with large antennae. They have an armored exoskeleton and are distinguished by a single median eye on a transparent head. Copepods have two pairs of antennae with the first pair being conspicuous and long.

The body is divided into a head, thorax and trunk segments. The head is fused with the first thoracic segments. On these segments are modified appendages, which help with feeding. They have short cylindrical bodies with the anterior segments bearing swimming appendages ending in a pair of caudal rami at the base of the abdomen.

Copepods are important to global ecology and to the carbon cycle. They are dominant members of zooplankton and a food source for small fish and krill. They are the largest animal biomass on earth competing only with krill. Their small size and fast growth rates coupled with the fact that they are evenly distributed throughout the world's oceans contribute hugely to secondary productivity in our oceans and the global ocean carbon sink. Surface layers of the oceans are believed to be the world's biggest carbon sink, absorbing up to 2 billion tonnes of carbon a year. Copepods sink to deeper waters at night to avoid predators. Their molted exoskeletons and faecal pellets bring carbon away from the surface of the ocean to the deep sea.

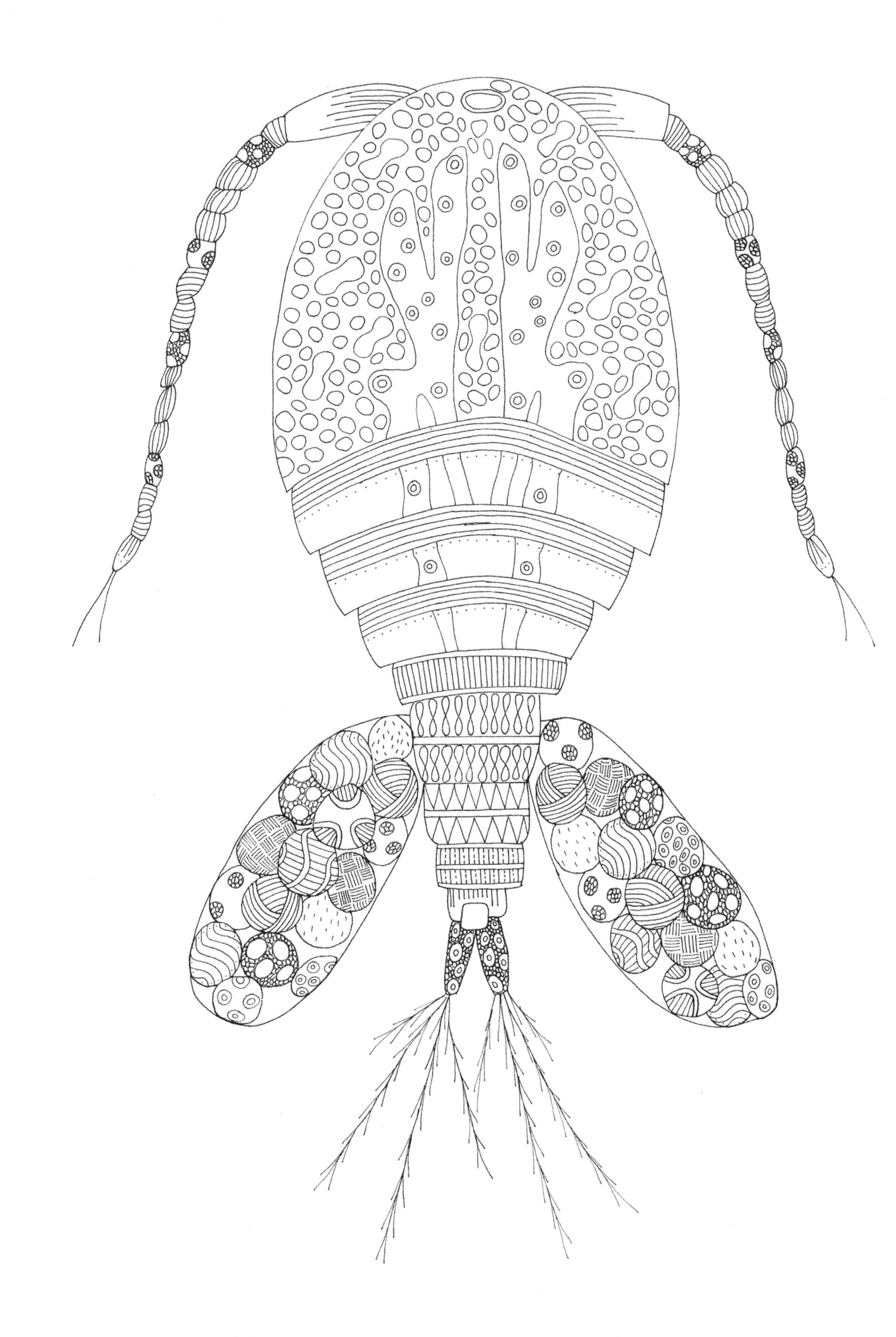

Shrimp zoea

The zoea is the common larva of decapods and has variations in its features in different species. It is characterised by a large cephalothorax covered in a helmet-like carapace.

The abdomen is split into six segments and has no appendages. It terminates in a caudal furca and telson which are used to move in the water column.

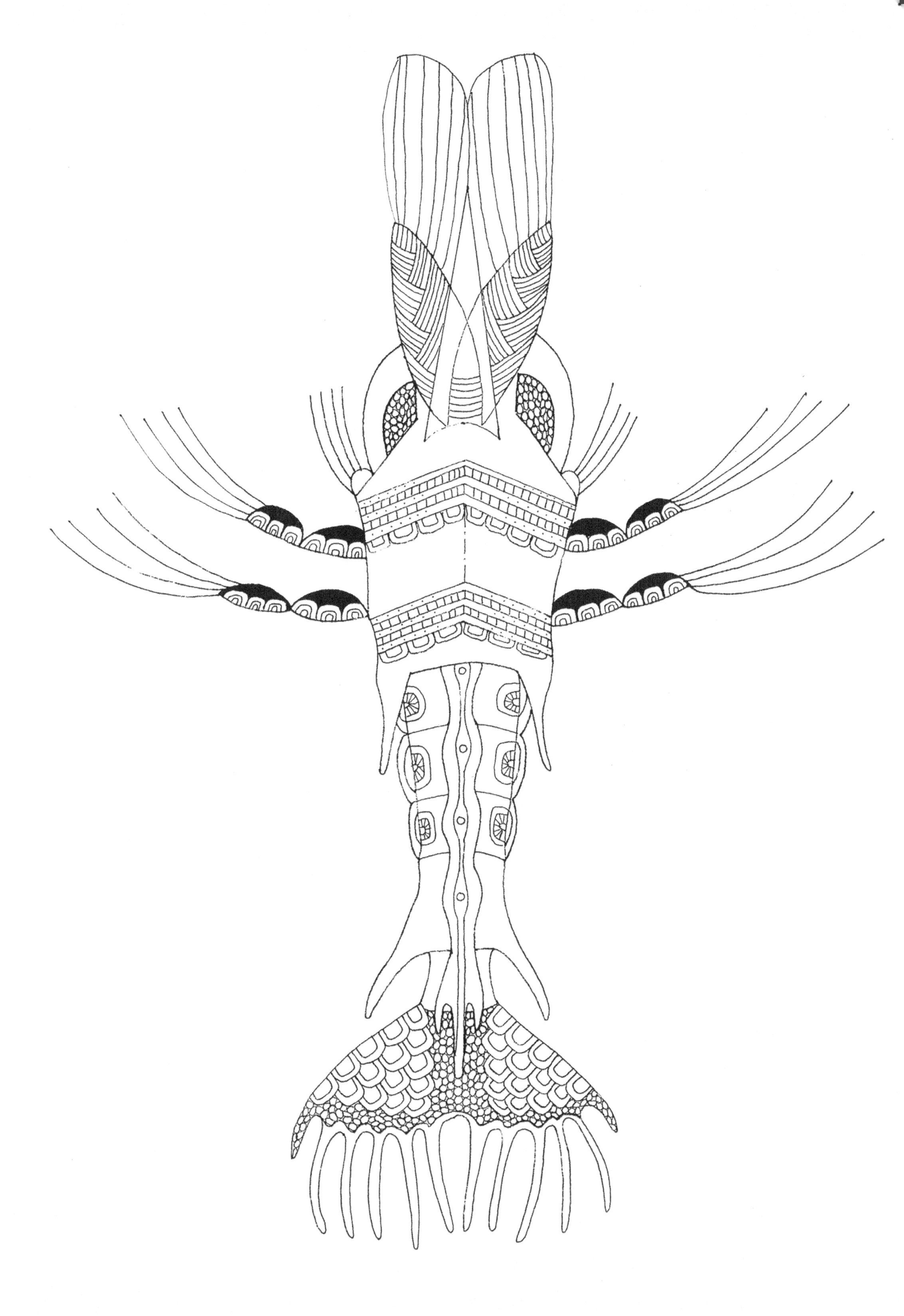

Shrimp zoea

The shrimp zoea has a rostrum at the front and a large pair of compound eyes. Its antennules and antennae are short and sensory in function. Swimming is made possible by the first and second maxillipeds.

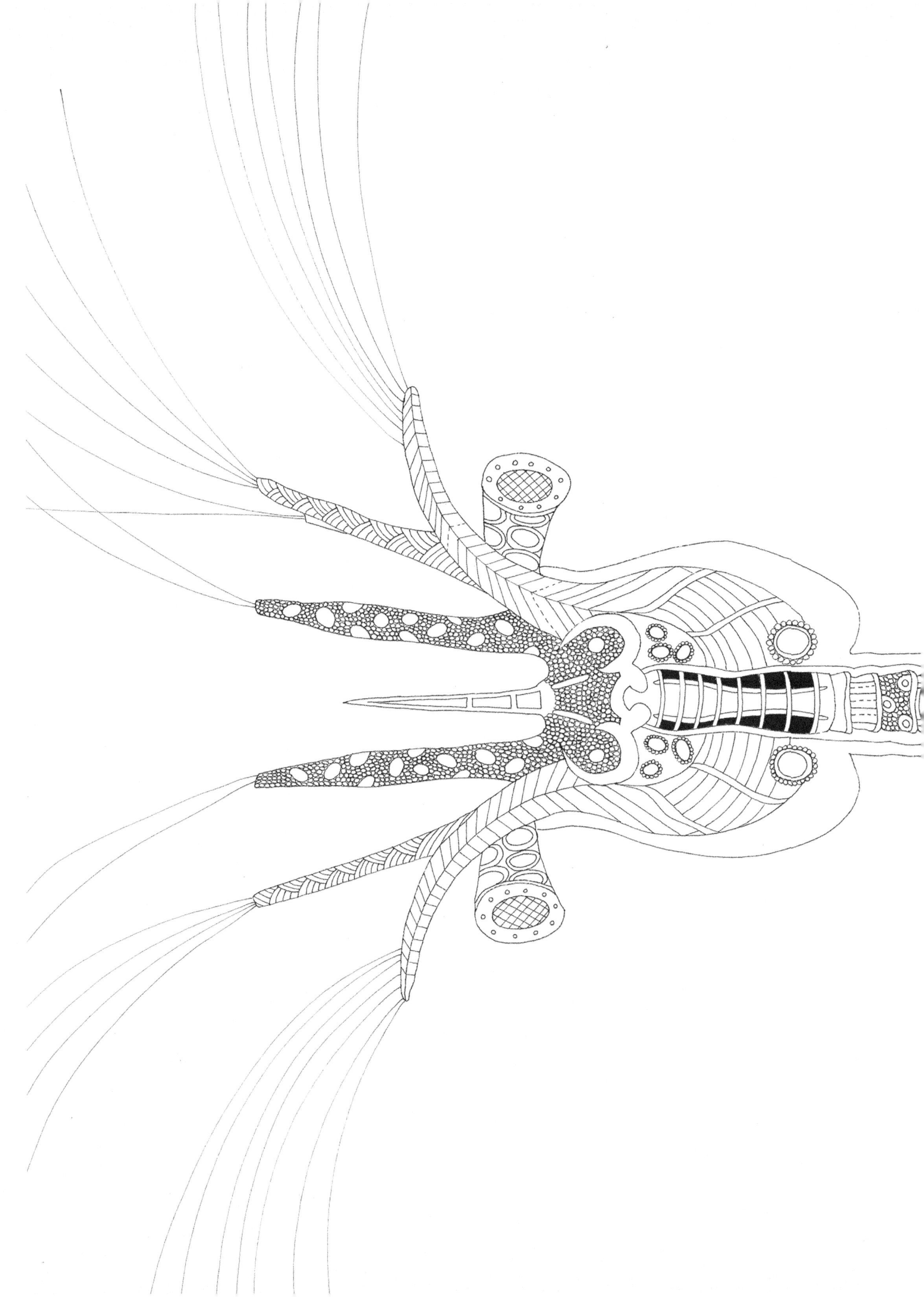

Honey Bee mouthparts

Honey bees belong to the *Apis* genius of insects and are among 20,000 species of bee that exist today. They co-evolved with flowering plants some 130 million years ago in South and Southeast Asia and today they can be found in every pocket of the earth. They play a vital role in sustaining the planet's ecosystem.

Honey bees have highly adapted mouthparts that enable them to collect nectar and pollen from flowers. Their short mandibles are suited to molding wax, chewing wood, gathering pollen and biting. The labium which forms part of the proboscis is made up of a segmented palp that is sensory and curves to form a tube that the bee uses to suck nectar. Honey is made from this nectar and sweet deposits that are gathered from trees and plants and is then modified and stored in honeycomb. This honey can sustain an entire colony of honey bees.

The head houses the visual, olfactory and taste centres. Two antennae have thousands of sensory organs which are specialised for taste, touch and smell. They "hear" sound that is close to them by sensing air movement in their mechanoreceptors. Three simple eyes, located at the tip of the head behave as light sensors, while the large compound eyes have up to 8000 cells which are highly evolved to detect motion.

Honey bees are regarded as guardians of the food chain. Not only are a third of the food that we consume pollinated by honey bees, but they also pollinate the seeds, fruit and berries eaten by birds and small mammals.

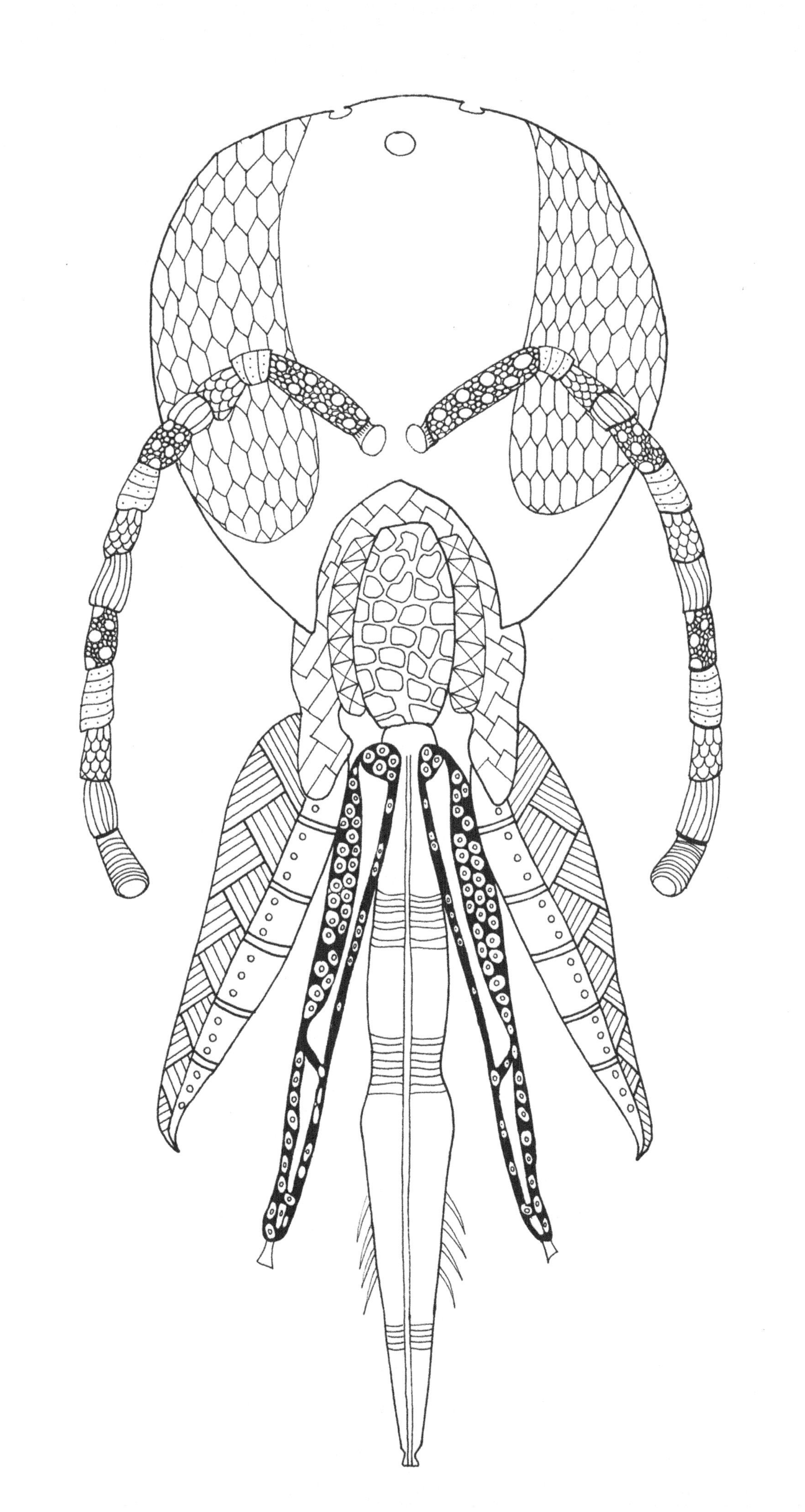

Wasp head

Wasps belong to the order Hymenotera and are found in all parts of the world except the polar regions. They have a hard exoskeleton which is used for protection and their bodies are divided into a head, mesosoma and metasoma. The wasp's head has two large compound eyes and two antennae used for sensing and smelling. It possesses biting and cutting mandibles and a proboscis which helps them drink nectar.

Wasps are beneficial to humans as nearly every pest insect on Earth is preyed upon by a wasp species, as they provide a host for their parasitic larvae.

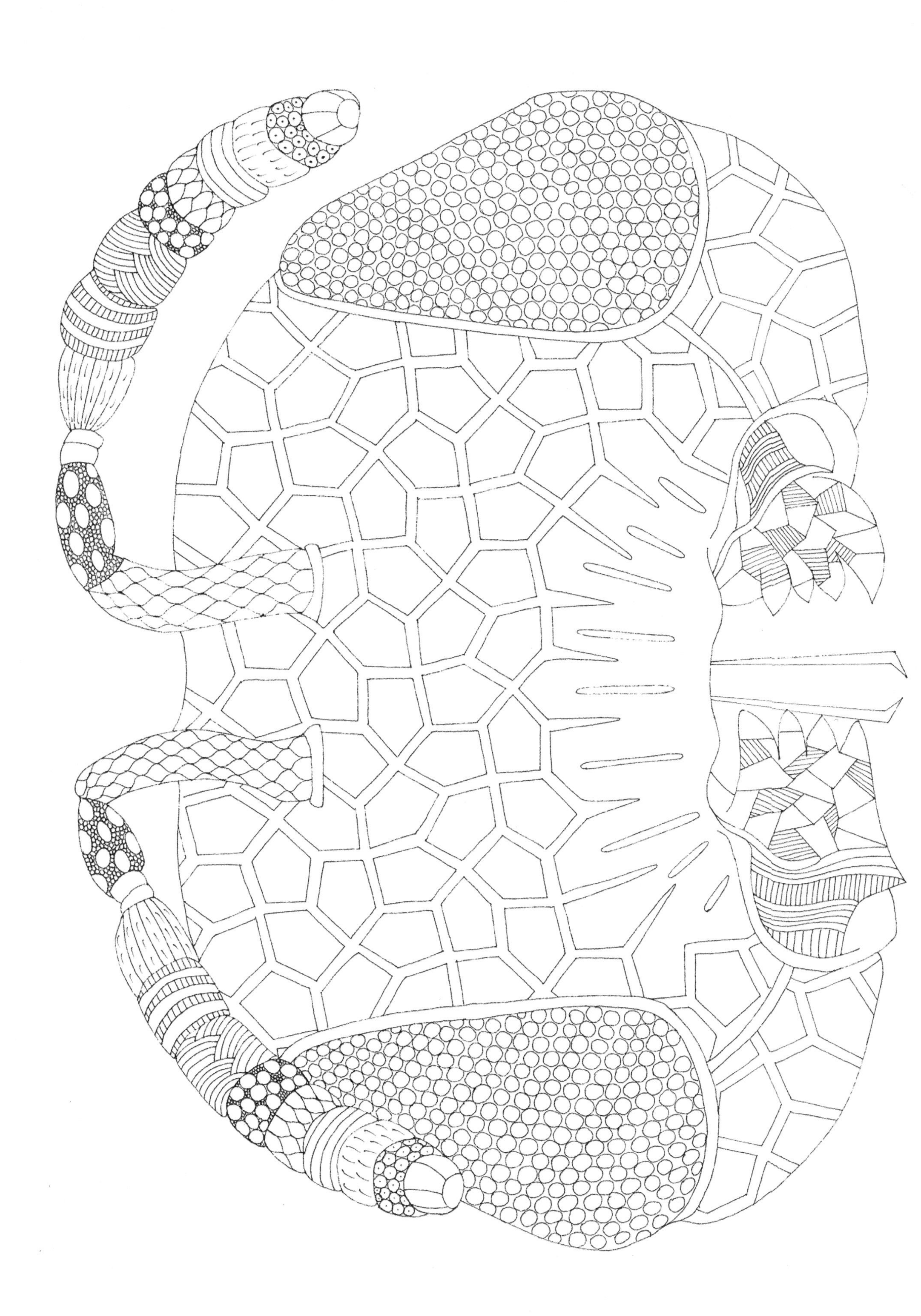

Mosquito pupa

Mosquitos go through four distinct stages in their lifecycles. Eggs develop into larvae, then onto a pupa and finally an adult mosquito.

The pupa is the resting, non-feeding stage of development. When viewed on its side the pupa resembles a comma. The head and thorax are merged to form a cephalothorax, with an abdomen curling around underneath. It uses its abdomen to actively swim in the water column, avoiding predators and moving to the surface for air.

After a few days as a pupa, the underside of the cephalothorax splits open and the adult mosquito emerges.

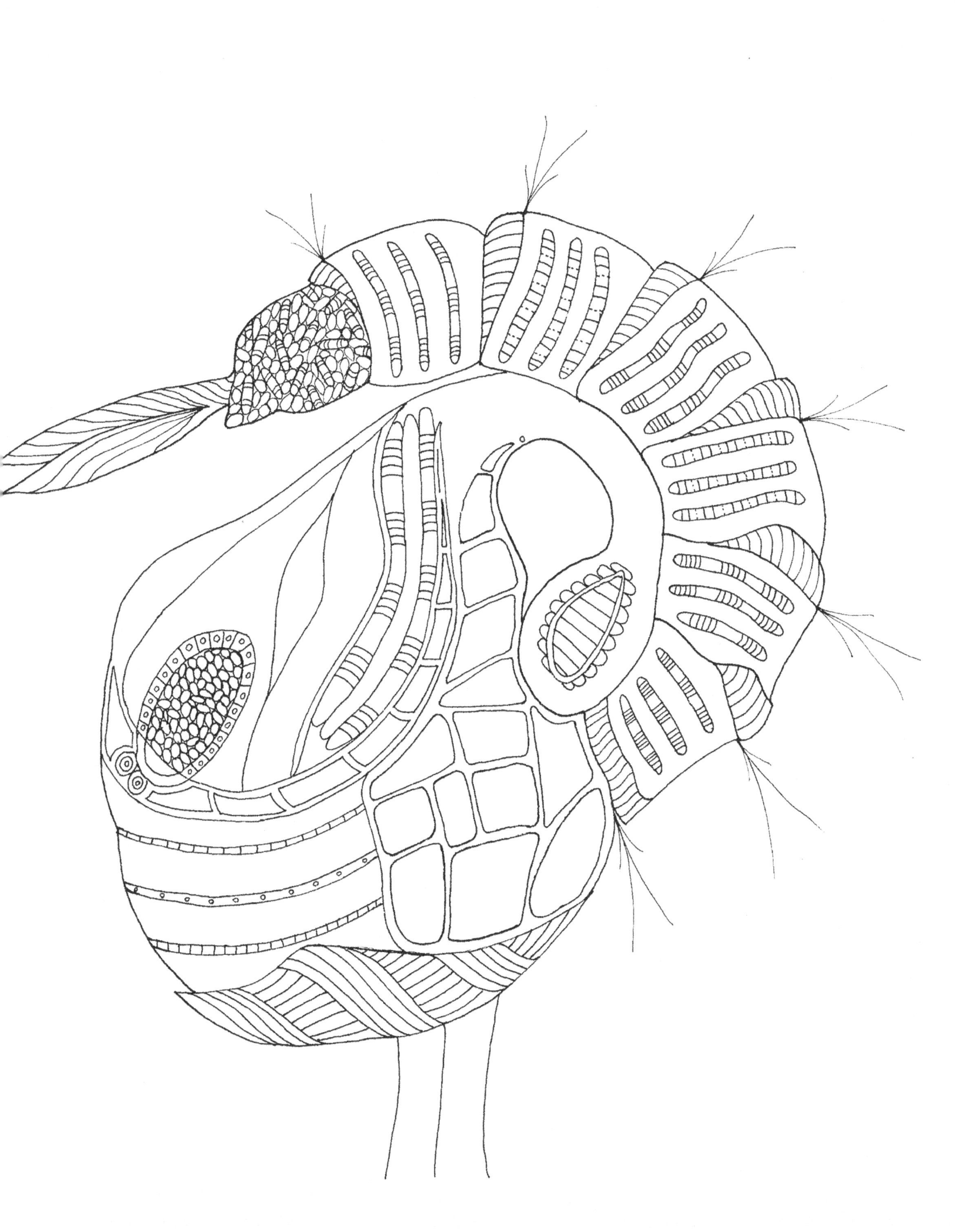

Cabbage butterfly eggs

Pieris brassicae, the white cabbage butterfly, live in large open spaces, agricultural areas, meadows and parks of Europe, North Africa and Asia. They lay their pale yellow eggs on cabbage leaves where they eventually hatch into caterpillars, getting nourishment from the leaves on which they are laid. Many eggs are laid during their short lifespan to ensure that even a small number of eggs actually survive.

Cabbage butterflies lay their eggs in batches. They are attached to the underside of the leaf of the cabbage plant by secreting a fast drying glue-like chemical.

Butterfly eggs vary in shape, size and colour. Some are spherical, pod-shaped or oval and can be coloured white, yellow and green. The eggs shell is decorated with raised ribs or pits and has a micropyle, which is a small pit on top which marks where sperm enters the egg.

Fertilised eggs mature within the female butterfly for around 3-6 days before they are laid. Their pale yellow colour eventually turns a darker shade of yellow within 24 hours of being laid. Just before they hatch the shell turns from black to transparent and the larvae then appears. It gnaws open the egg and then eats the remaining egg for its first meal.

Cabbage butterfly eggs are threatened by birds, amphibians, reptiles and grazing mammals. They are also prone to parasitisation by wasps and flies. In many places these butterfly eggs and larvae are unwanted pests due to their potential effect on crops.

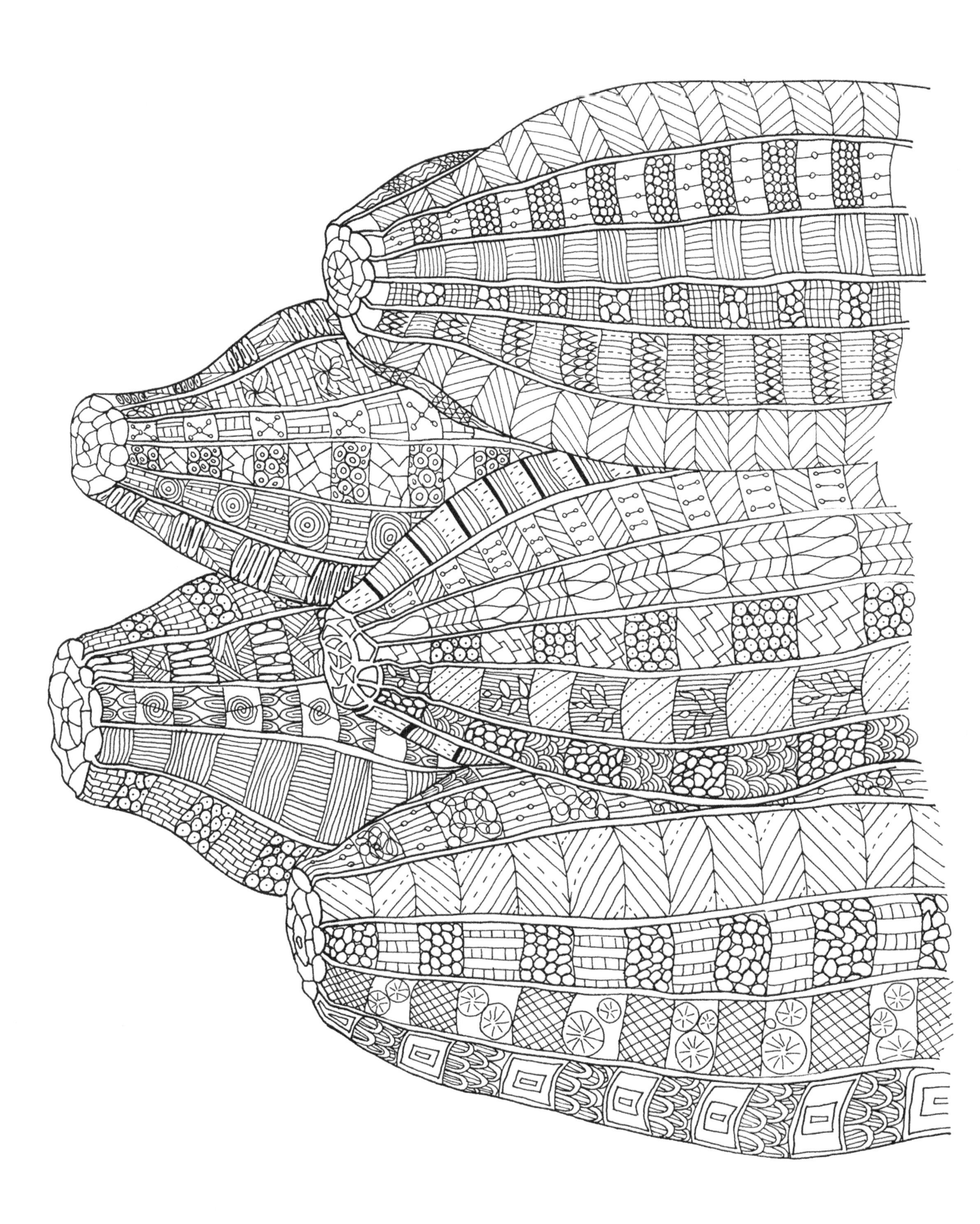

Sea angel

Sea angels are predatory sea slugs. They live as plankton from the Arctic to the Antarctic and feed on sea butterflies and other zooplankton. They use tentacles to grasp their prey and swim slowly using a foot which is modified into two wing-like parapodia.

Clione limacine, illustrated here, can grow up to 5cm.

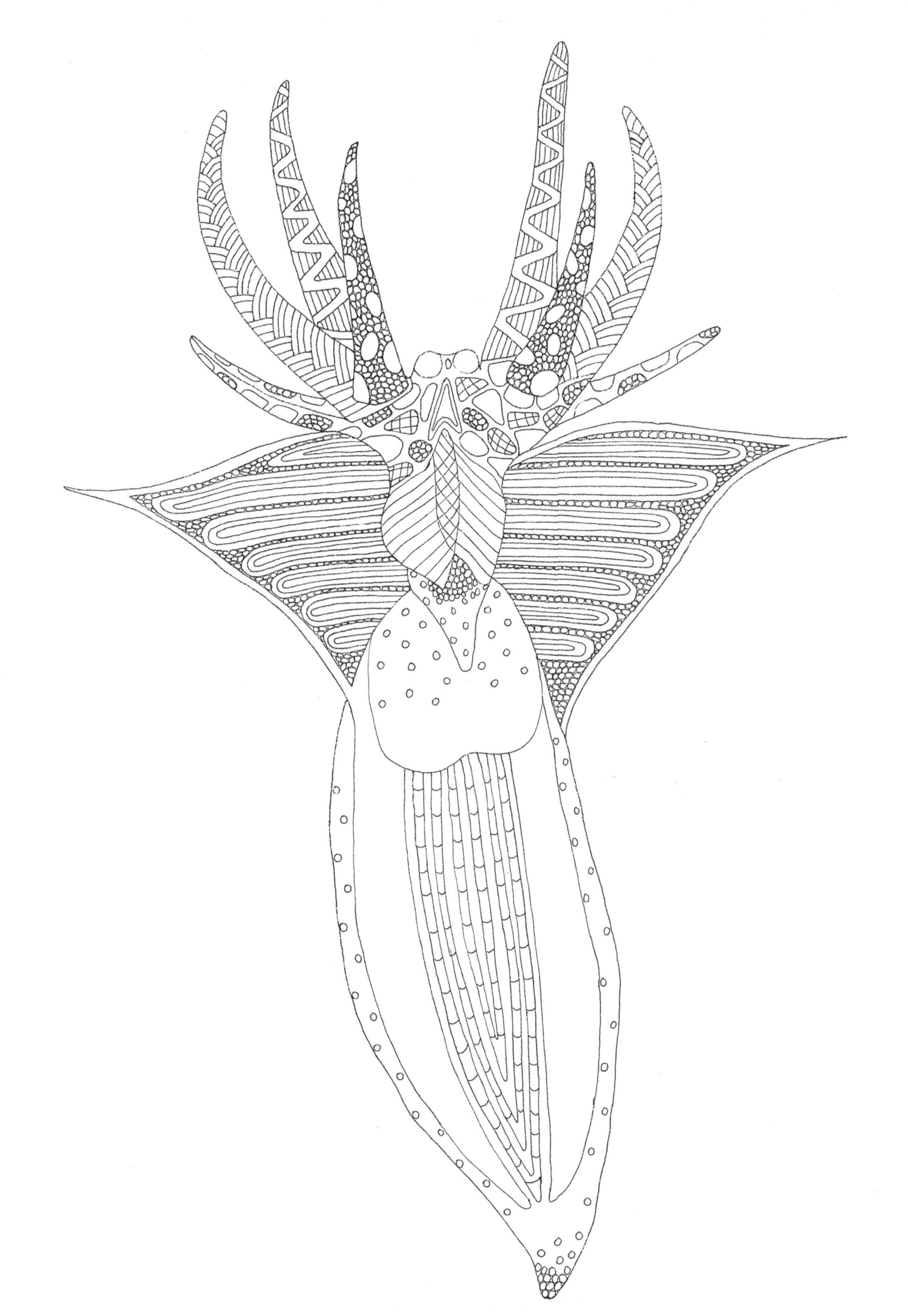

Sea urchin larva

Sea urchins are Echinoderms, a phylum that includes starfish, sea cucumbers and crinoids. Their larvae form part of the merozooplankton, animals which complete only one part of their life cycle in the plankton.

These larvae begin with a small number of long arms which eventually increase in number. Their arms have bands of cilia which are used in movement and feeding. Within the arms exists an internal skeleton of calcium carbonate. They have a mouth on top with a central gut, terminating in an anus.

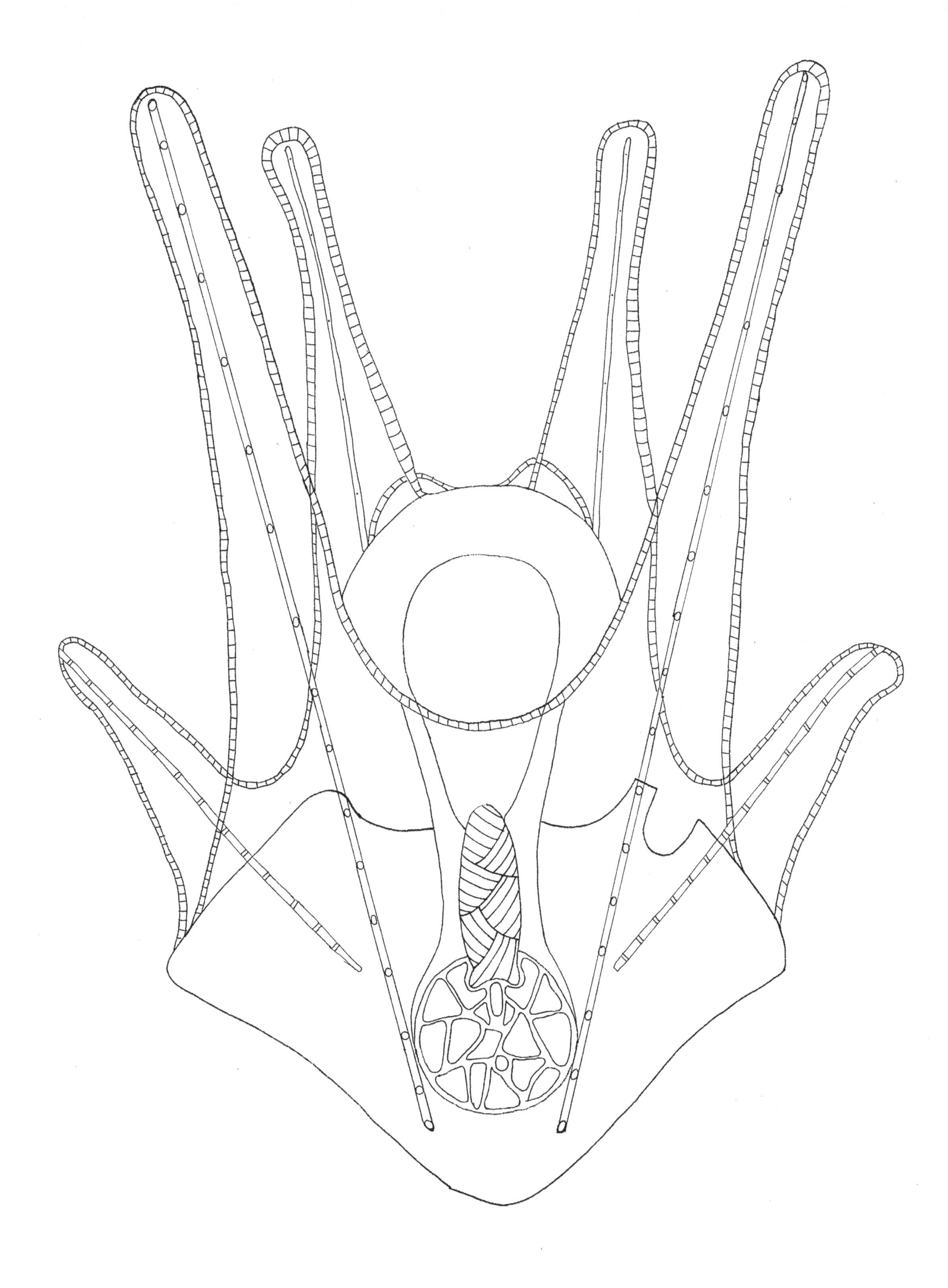

Jennifer Delaney has spent 15 years teaching Mathematics in rural Donegal, Ireland. She has a background in Marine Science and is a self-trained illustrator. She uses both art and science as a means to investigate and make sense of the world in which we live. She shares her love of art, science and nature with her husband and four young children.

Please share your finished coloured artwork on Twitter, Facebook and Instagram with @jvcdelaney #lifeunderthelens

Made in the USA
Middletown, DE
31 August 2021